T0220302

Cambridge Elements ⚌

Elements in Decision Theory and Philosophy
edited by
Martin Peterson
Texas A&M University

RATIONAL CHOICE USING IMPRECISE PROBABILITIES AND UTILITIES

Paul Weirich
University of Missouri

CAMBRIDGE
UNIVERSITY PRESS

CAMBRIDGE
UNIVERSITY PRESS

University Printing House, Cambridge CB2 8BS, United Kingdom

One Liberty Plaza, 20th Floor, New York, NY 10006, USA

477 Williamstown Road, Port Melbourne, VIC 3207, Australia

314–321, 3rd Floor, Plot 3, Splendor Forum, Jasola District Centre, New Delhi – 110025, India

79 Anson Road, #06–04/06, Singapore 079906

Cambridge University Press is part of the University of Cambridge.

It furthers the University's mission by disseminating knowledge in the pursuit of education, learning, and research at the highest international levels of excellence.

www.cambridge.org
Information on this title: www.cambridge.org/9781108713504
DOI: 10.1017/9781108582209

© Paul Weirich 2021

First published 2021

A catalogue record for this publication is available from the British Library.

ISBN 978-1-108-71350-4 Paperback
ISSN 2517-4827 (online)
ISSN 2517-4819 (print)

Rational Choice Using Imprecise Probabilities and Utilities

Elements in Decision Theory and Philosophy

DOI: 10.1017/9781108582209
First published online: February 2021

Paul Weirich
University of Missouri

Author for correspondence: Paul Weirich, WeirichP@missouri.edu

Abstract: An agent often does not have precise probabilities or utilities to guide resolution of a decision problem. I advance a principle of rationality for making decisions in such cases. To begin, I represent the doxastic and conative state of an agent with a set of pairs of a probability assignment and a utility assignment. Then I support a decision principle that allows any act that maximizes expected utility according to some pair of assignments in the set. Assuming that computation of an option's expected utility uses comprehensive possible outcomes that include the option's risk, no consideration supports a stricter requirement.

Keywords: decision theory, game theory, imprecise probabilities, sequences of choices, imprecise utilities

ISBNs: 9781108713504 (PB), 9781108582209 (OC)
ISSNs: 2517-4827 (online), 2517-4819 (print)

Contents

1 Introduction

A college graduate deciding whether to enter medical school may not assign precise probabilities or precise utilities to the possible outcomes of her options. She may not evaluate precisely her chances of succeeding in medical school or the attraction of life as a physician. A person making a choice in a decision problem often does not have enough information or experience to assign precise probabilities and utilities to her options' possible outcomes. Common principles of rational choice, such as the principle to maximize expected utility, cover only decision problems with precise probabilities and utilities for possible outcomes. However, a general account of rational choice must also cover decision problems with imprecise probabilities and utilities for possible outcomes.

An explanation of imprecise probabilities and utilities includes as a foundation a thorough account of precise probabilities and utilities. I take probabilities as rational degrees of belief and utilities as rational degrees of desire and in support of this interpretation argue that rational degrees of belief comply with the axioms of probability and that rational degrees of desire comply with the principle of expected utility. Then I extend this view to imprecise probabilities and utilities. Afterwards, I advance a decision principle that uses imprecise probabilities and utilities to identify rational choices.

The decision principle applies within a model of choice that makes several idealizations about agents and their decision problems. The principle lays the groundwork for more general principles that dispense with the idealizations.

My treatment of imprecision briefly discusses rival positions but does not thoroughly review the literature. References direct readers to alternative stances.

Section 2 characterizes imprecise probabilities and utilities and Section 3 defends their rationality. Sections 4 and 5 present constraints that rationality imposes on precise probabilities and utilities and Section 6 extends the constraints to imprecise probabilities and utilities. Section 7 formulates a principle of rational choice that accommodates imprecise probabilities and utilities. Section 8 applies the principle in sequences of choices and Section 9 applies the principle to choices in games of strategy. Section 10, the final section, draws conclusions.

2 Imprecision

An account of imprecise probabilities and utilities grows out of an account of precise probabilities and utilities. This section explains what probabilities and utilities are and then what it means for probabilities and utilities to be imprecise. Later sections explain how to use imprecise probabilities and utilities to make decisions. The principle of rational choice I present relies on this section's points about agents, their decision problems, and their resources for resolving their decision problems.

2.1 A Decision Model

I advance a principle of rational choice in a decision model – that is, under a set of assumptions about agents and their decision problems. The principle uses an agent's belief states, or doxastic states, and the agent's desire states, or conative states. The model idealizes agents but assumes that they have doxastic and conative states of the type that humans have. Psychology describes the types of mental states that humans have and philosophical points about mental states help define and individuate them. I use a lay understanding of psychology to describe doxastic and conative states, attending especially to features philosophically important for the model's principle of rational choice.

The decision model incorporates several idealizations and simplifying assumptions about agents, their circumstances, and their decision problems. In the model, agents are cognitively ideal and so know all a priori truths they entertain – but need not entertain every a priori truth even though reflection is effortless and instantaneous for them. Furthermore, ideal agents know their own minds and so are aware of their doxastic and conative states and their cognitive powers. I assume that, in the decision problems they face, they are aware of the relevant characteristics of the problems and resolve the problems without distraction. They frame their decision problems using sentences that express their options and the possible outcomes of their options. I assume that they have information sufficient to understand fully the proposition any sentence expresses (in the language they use, with at least the expressive power of English). Therefore, they recognize when two sentences express the same option or the same possible outcome of an option.

Lacking evidence concerning a proposition's truth, an ideal agent need not assign a precise probability to the proposition, even given unlimited reflection. An agent's reflection on the proposition and the grounds of a probability assignment to the proposition cannot make up for a lack of information. An imprecise probability may remain imprecise, despite reflection, if new information does not arrive. Similarly, a utility that is imprecise because of a lack of experience may remain imprecise, despite reflection, if new experience does not arrive. Reflection cannot make up for a lack of experience.

A probability is imprecise if no single number accurately characterizes it. In this case, a number characterizing it is indeterminate. The same holds for an imprecise utility. Its imprecision entails the indeterminacy of a number characterizing the utility.

Because a proposition's probability is a number representing an agent's doxastic attitude to the proposition, its indeterminacy characterizes a representation of the attitude and not the attitude itself. The attitude has exactly the features it has. The same point applies to the indeterminacy of a proposition's utility. A proposition's utility is a number representing an agent's conative attitude to the proposition and its indeterminacy characterizes a representation of the agent's conative attitude. The attitude, given exactly its features, has an indeterminate numerical representation. Moreover, even if a doxastic or conative attitude has an indeterminate representation by a single number, an alternative type of representation may accurately characterize the attitude.

I say that an attitude is *quantitative* if it has an apt representation that uses a single quantity, such as a number. If an attitude is nonquantitative, an apt representation may still use numbers, but not a single number, to represent the attitude. A representation of the attitude that uses an interval of numbers is quantitative although it does not specify a single number to represent the attitude. It imprecisely specifies a number to represent the attitude.

When evaluating a proposed principle of rationality, I assume that an agent in the decision model is rational in all respects except possibly for compliance with the principle and then consider whether rationality requires compliance with the principle. An argument for the principle rescinds the idealization that the agent is rational in matters the principle governs. Then it maintains that an agent's violation of the principle is irrational.

I take a *doxastic domain* for an agent to be a set of propositions to which the agent has doxastic attitudes.[1] These attitudes may yield probability assignments but may be just judgments of epistemic possibility or impossibility. The set is usually taken to be a *Boolean algebra* formed from a set of atomic propositions by closure under negation and disjunction. However, to allow for multiple equivalent propositions differing in structure, I take the set to form a *propositional language* constructed from a set of atomic propositions by closure under the standard propositional operations.[2]

---

[1] An agent's adopting a doxastic domain for probability assignments prevents inconsistencies that may arise if she assigns probabilities to all propositions.

[2] An agent may fail to assign probabilities to propositions in the doxastic domain she uses not because of imprecision but because she regards some propositions as infinitesimally less probable than others. I put aside such cases by assuming that the agent uses a doxastic domain that is *Archimedean* in the sense that, for any two epistemically possible propositions compared probabilistically, there is a natural number n such that one proposition is no more than n times more probable than the other.

In an agent's decision problem, the number of *salient* options is finite just in case in some adequate representation of the decision problem the options form a finite set such that the agent is not indifferent between any two options belonging to the set. In this case, a unique option maximizes utility; ties do not arise. A representation of the options in a decision problem may combine tying options, options such that the agent is indifferent among them, into a single option; the option may be the disjunction of the tying options. Realizing a disjunctive option is equivalent to realizing a disjunct.

Using this terminology, the decision problems that the decision model treats have the following characteristics. The number of salient options is finite and options have a finite number of possible outcomes with probabilities and finite, stable utilities not altered by an option's adoption. If the probabilities and utilities of possible outcomes are imprecise, they nonetheless have an adequate representation.

An agent's probability assignments may use various doxastic domains. In the decision model, suppose that an agent adopts a doxastic domain for a decision problem, assigns probabilities to propositions in the domain that express possible outcomes of options, and reaches a choice. The agent's adopting another doxastic domain for the decision problem does not reverse the choice. Each doxastic domain yields the same choice. So, although a doxastic domain's selection is arbitrary, its arbitrariness is inconsequential. An agent's doxastic domains differ in the events to which they assign probabilities but do not differ, for propositions to which they assign probabilities, in features that affect choice.

If a rational ideal agent ideally situated in a decision problem that the decision model treats assigns precise probabilities and utilities to the possible outcomes of options, an option is rational if and only if it maximizes expected utility. Section 7 generalizes this principle to cover decision problems in the model without precise probabilities and utilities. A further extension of the principle beyond the model may follow common methods of generalizing a principle in a model by relaxing idealizations and revising the principle to accommodate new situations.

2.2 Probability

Probabilities come in two sorts. *Physical probabilities* arise from physical features of events, such as the physical features of a coin toss. *Evidential probabilities* are relative to evidence and an agent's evidential probabilities are relative to the agent's evidence. Imagine a courtroom trial of a defendant for commission of a crime. The physical probability that the defendant is guilty depends on the defendant's past acts and is either 0 or 1, depending on whether the defendant committed the crime. For a juror, the evidential probability that the accused is guilty depends on the evidence presented during the trial and may rise or fall as evidence is presented even though the physical probability of guilt is constant. At the end of the trial, the evidential probability of guilt may have a non-extreme value although the physical probability of guilt remains either 0 or 1. Physical probabilities are sometimes called objective but probabilities may be objective in many ways. For example, an evidential probability may be objectively settled by the evidence. Evidential probabilities are sometimes called epistemic but they are epistemic in a specific way; they are relative to evidence.

A rational ideal agent knows the evidential probabilities of an option's possible outcomes relative to her evidence; and her degrees of belief, which direct her choices, equal these evidential probabilities. Principles of rational choice using probabilities use evidential probabilities because an agent often does not know the physical probabilities of an option's possible outcomes. Because my topic is rational choice, by probability I generally mean evidential probability.

An evidential probability attaches to a proposition, the content of a declarative sentence. I take a proposition to have a structure similar to the structure of a sentence. Two logical truths may therefore have different structures and so be different propositions, although each is true in all possible worlds. Because two distinct propositions may be true in exactly the same possible worlds, a set of possible worlds does not adequately represent a proposition. Sentences and the propositions they express are similar, so I sometimes speak of the probability of a sentence, meaning the probability of the proposition that the sentence expresses.

I understand events in a technical sense that includes states. Propositions represent events such as acts, states of the world that settle the consequences of acts, and the outcomes of acts. Probabilities and utilities attach to events by attaching to propositional representations of the events.[3]

The term imprecise probability is a bit misleading because a probability in the ordinary sense is precise. Indeterminate probability is a more suggestive term.[4] However, I use the term imprecise probability, taking it in a technical sense that does not entail being a probability, because this usage is widespread. An imprecise probability is an imprecise specification of a probability, such as an interval of probabilities.[5]

---

[3] Jeffrey ([1965] 1990) attaches probability and utilities (desirabilities) to propositions but takes propositions to be adequately represented by sets of possible worlds.

[4] Levi (1974) recommends using the term indeterminate probability.

[5] Walley (1991) provides a classic account of imprecise probabilities. Bradley (2019) and Mahanti (2019) offer recent surveys. Augustin et al. (2014), Troffaes and de Cooman (2014), and Zaffalon and Miranda (2018) present mathematical results.

A person may assign probabilities to some, but not all, propositions of a doxastic domain she adopts. Her probability that heads turns up on a coin toss may be precisely 50 percent. However, because of sparse information, she may not assign a precise probability to rain tomorrow. Imprecision concerning the atoms of a doxastic domain may spread to compounds formed from the atoms. For example, imprecise probabilities for two atoms may lead to an imprecise probability for their disjunction.

Familiar doxastic states include belief, suspension of judgment, and disbelief. A *degree of belief* quantitatively represents a doxastic state. The relation between belief and degree of belief clarifies degree of belief. However, the relation is subtle. A belief is not simply a high degree of belief, as the threshold for what counts as a high degree of belief must at least vary with context to accommodate beliefs about lottery tickets – an agent typically has a very high degree of belief that any given lottery ticket will lose but still does not believe that the ticket will lose and instead suspends judgment. The doxastic attitudes that degrees of belief represent explain beliefs but in a complex way. Assuming that an agent is both cognitively ideal and rational simplifies an account of the relation between belief and degree of belief because the assumption puts aside cases in which, for example, an agent irrationally believes a proposition to which she assigns a low degree of belief. However, even for a rational ideal agent, the relation between belief and degree of belief is intricate. I do not describe the relation except to say that belief that a proposition is true is generally the product of a high degree of belief that the proposition is true, according to a context-sensitive threshold for being high. I do not need to be more specific about the relation because I treat in detail only degree of belief, and not belief, and so do not need a detailed, unified account of doxastic attitudes.

I take a doxastic attitude that a degree of belief represents as a primarily passive response to evidence (such as observation) and not as a primarily active representation of the world (so that it is evaluable as an act). The doxastic attitude is sometimes called a strength of belief, with the understanding that minimum strength is not belief but disbelief. An agent's degree of belief that one proposition holds is greater than the agent's degree of belief that another proposition holds only if the agent believes the first proposition more strongly than the second proposition, again with the understanding that, when the two degrees of belief are low, the agent typically does not believe either proposition but instead disbelieves both. Degree of belief has a technical sense according to which it does not measure only belief but also disbelief. Some authors use the term credence instead of degree of belief to disavow a restriction to belief.

Degrees of belief represent doxastic attitudes. A degree of belief that a proposition holds is a number representing an agent's doxastic attitude to the proposition. The number represents strength of belief (in a technical sense that includes strength of disbelief), with, by convention, 1 representing maximum strength and 0 representing minimum strength. Because degrees of belief are numbers, they can comply with the laws of probability. I take evidential probabilities as rational degrees of belief.

Representations of attitudes differ from the attitudes themselves. A degree of belief is a number and not itself a doxastic attitude. For brevity of expression, theorists sometimes speak of a degree of belief as if it were the attitude it represents. They say, for example, that a person's high degree of belief that a proposition is true explains the person's belief that it is true. Strictly speaking, they mean that the doxastic attitude, the strength of belief, that the high degree of belief represents explains the person's belief. For convenience, I also sometimes speak of a degree of belief as if it were a doxastic attitude, although strictly speaking it is a number representing a doxastic attitude.

By convention, degrees of belief, as I understand them, represent ratios of strengths of belief not just differences in strengths of belief. Hence, degrees of belief use a ratio scale rather than, say, an interval scale, so that, if an agent's degree of belief that p is 0.6 and the agent's degree of belief that q is 0.3, then the agent believes p twice as strongly as q.

An account of degree of belief may implicitly define it by advancing principles governing it that are sufficient for grasping the meaning of degree of belief. Probabilism, the view that rational degrees of belief comply with the laws of probability, a view that Section 4 advocates, contributes to an account of degree of belief that implicitly defines it using, among other principles, the laws of probability as norms for degrees of belief. Psychological descriptions of the causes of strengths of belief, such as evidence, and the effects of strengths of belief, such as acts, further supplement this section's brief introduction of degrees of belief.

2.3 Utility

Conative attitudes include desire, indifference, and aversion. The conative attitudes that degrees of desire represent are quantitative counterparts of desire, indifference, and aversion. Although an agent may form a conative attitude, such as a desire, putting aside some considerations, degrees of desire represent attitudes formed all things considered.

As is degree of belief, degree of desire is a technical term. Using indifference as a zero point for a scale of degree of desire, a negative degree of desire represents an aversion and a positive degree of desire represents a desire.[6]

---

[6] In one sense, indifference holds between two items; an agent is indifferent between them. In another sense, indifference is toward a single item. An agent may be indifferent to dessert.

Accordingly, the relation between desire and degree of desire is less complex than the relation between belief and degree of belief.

A desire and a conative attitude represented by a degree of desire are distinct attitudes but may have the same realization. The same mental state may be classified as a quantitative attitude that a degree of desire represents and also classified as a nonquantitative desire, aversion, or attitude of indifference.

Desire and degree of desire and, similarly, aversion and degree of aversion (or negative degree of desire) are passive attitudes responding to events entertained (and are not evaluable as acts), although they prompt acts to satisfy desires and to prevent realizations of aversions. As for degree of belief, degree of desire has an implicit definition given by principles governing it, including the normative principle of expected utility that requires an act's degree of desire to equal the expected degree of desire of the act's outcome. An account of the causes of strengths of desire, such as envy, and the effects of strengths of desire, such as acts, fills out the implicit definition.

The conative attitudes that degrees of desire represent are propositional attitudes, that is, attitudes directed toward propositions, as are desire, indifference, and aversion. Because degrees of belief and degrees of desire alike attach to propositions, principles such as the expected-utility principle may join them seamlessly.

Degrees of desire that satisfy the laws of utility are called utilities, just as degrees of belief that satisfy the laws of probability are called probabilities. Utilities as well as probabilities may be imprecise. A person may assign a precise utility to gaining a thousand dollars but not assign a precise utility to holding a lottery ticket that if drawn yields a thousand dollars because she does not know the number of tickets in the lottery. Also, a person who lacks the experience of eating passion fruit may not assign a precise utility to eating this fruit. A conative attitude, such as a desire, may have an indeterminate numerical representation because of a lack of information or a lack of experience. For consistency of terminology, just as I use the term imprecise probability instead of indeterminate probability, I use the term imprecise utility instead of indeterminate utility. An imprecise utility is an imprecise specification of a utility, such as an interval of utilities.

Theorists often claim that an agent, even with reflection, may not have a preference, or be indifferent, between two events, such as hiking along Hurricane Ridge in the Olympic National Park and listening to a performance of Beethoven's Ninth Symphony by the Chicago Symphony Orchestra. She may

---

Indifference as a zero point is indifference in the second, non-relational sense. The two senses are closely related. An agent is indifferent to dessert if and only if she is indifferent between having dessert and not having it.

not be able to compare the events along any convenient dimension of comparison, not even pleasure, because the pleasures the events generate are of different types. If a rational ideal agent assigned both the hiking and the listening a precise degree of desire, then she would prefer the event with the higher degree of desire, and if their degrees of desire were the same would be indifferent between them. Because degrees of desire entail comparability, incomparability entails the absence of degrees of desire and so imprecision.[7]

For an ideal agent, degrees of desire, when rational, comply with the laws of utility, such as the expected-utility principle. Therefore, I take utilities as rational degrees of desire. When a rational ideal agent in a decision problem fails to assign degrees of desire to her options, the options have imprecise utilities for her.

2.4 Constructivism

My account of degrees of belief takes them to represent doxastic attitudes to propositions. It counts as a *realist* view, as opposed to a *constructivist* view that takes degrees of belief to represent not attitudes but choices or preferences and so to be constructed from choices or preferences rather than to have an independent reality. One constructivist account, following de Finetti ([1937] 1964), defines an agent's degree of belief that a proposition holds as the smallest percent of a dollar that the agent will exchange for a bet that gains a dollar if the proposition holds and otherwise nothing. The norm of expected-utility maximization, assuming that dollar amounts equal utilities, requires that the agent pay for the bet a percent of a dollar no greater than the agent's degree of belief that the proposition holds. This constructivist account makes the relation between the degree of belief and the exchange rate hold by definition, whereas my realist account accommodates the relation's being a normative requirement.

Another constructivist account defines an agent's degree of belief that a proposition holds as the value of the probability function that represents the agent's doxastic comparisons of propositions. This account assumes that the comparisons satisfy certain conditions, presented by Krantz and colleagues (1971: chap. 5), that ensure the existence and uniqueness of a probability function representing the comparisons. The definition makes having degrees of belief that satisfy the axioms of probability dependent on an arbitrary selection of a representation of doxastic comparisons, given that some perfectly adequate representations do not use a probability function, as, for example, Titelbaum (forthcoming: sec. 14.1) observes. In contrast, my realist account

---

[7] Chang (1997) offers a collection of essays on incomparability.

takes satisfying the axioms of probability as a normative requirement for degrees of belief. It also strengthens the norms for doxastic comparisons of propositions: not only must the comparisons be representable as agreeing with a probability function but they must also agree with the particular probability function that rational degrees of belief form.

A third constructivist account, for both degrees of belief and degrees of desire, defines an agent's degree of belief that a proposition holds as the value of the probability function that, along with a utility function, represents the agent's preferences among gambles as following expected utilities. This account assumes that the preferences satisfy certain conditions, for example those presented by Savage ([1954] 1972), that ensure the existence and uniqueness of the probability function and the existence and uniqueness (up to a positive linear transformation) of the utility function that together represent the preferences. The definition makes having preferences among gambles that follow expected utilities dependent on an arbitrary selection of a representation of the preferences, because some perfectly adequate representations of the preferences do not have them follow expected utilities, as, for example, Titelbaum (forthcoming: sec. 8.3) observes.[8] In contrast, my realist account takes following expected utilities as a normative requirement for preferences among gambles. The requirement is not just that preferences among gambles be representable as following expected utilities; they must follow expected utilities, calculated using degrees of belief and degrees of desire that are defined independently of preferences among gambles, as in Subsections 2.2 and 2.3. These degrees of belief and degrees of desire uniquely represent an ideal agent's doxastic and conative state, assuming it is quantitative, given a scale for degrees of desire.

Defining probabilities and utilities using choices, so that choices maximize expected utility by definition, destroys the power of the principle of expected-utility maximization to explain the rationality of choices. If choices maximize expected utility by definition, their maximizing expected utility cannot explain their rationality, not even their meeting requirements for having a representation as maximizing expected utility. Although the principle to choose as if maximizing expected utility does not use expected utility to explain the rationality of choices, the stronger, traditional decision principle of expected-utility maximization does, assuming an interpretation of probabilities and utilities according to which they represent propositional attitudes. It may take probability as rational degree of belief and utility as rational degree of desire, given that degrees of belief and degrees of desire represent propositional attitudes and exist independently of choices. The traditional principle of expected-utility

---

[8] Lyle Zynda (2000) and Meacham and Weisberg (2011) make similar observations.

maximization, because of its realist interpretation of probabilities and utilities, can explain the rationality of choices.

Constructivist accounts advance transitivity and other normative principles, governing doxastic comparisons of propositions and governing preferences among gambles, that are necessary for the representations they build. However, rationality advances stronger normative principles that require degrees of belief to conform to the probability axioms and that require preferences among gambles to follow expected utilities as computed using degrees of belief and degrees of desire representing propositional attitudes. The following sections argue for these stronger normative principles. They contribute to an explanation of the rationality of choices.

Although I put aside constructivist definitions of degrees of belief and degrees of desire, I retain methods of measuring degrees of belief and degrees of desire that use choices and preferences among gambles. Suppose that the preferences of a rational ideal agent have a unique representation (given a unit for utility) as following expected utilities. The probability and utility functions of the representation reveal, respectively, the agent's degrees of belief and degrees of desire because the agent's preferences follow expected utilities calculated using degrees of belief and degrees of desire that, being rational, form, respectively, probability and utility functions. In favorable conditions, degrees of belief and degrees of desire may be inferred from rational preferences among gambles because these preferences follow expected utilities.

A reluctance to acknowledge non-relational quantitative attitudes motivates replacing realist accounts of degrees of belief and degrees of desire, according to which they are, respectively, representations of doxastic and conative attitudes to propositions, with constructivist accounts of degrees of belief and degrees of degree, according to which they are just constructs representing preferences. However, I acknowledge non-relational quantitative attitudes. I take a single attitude on its own to be possibly quantitative, not quantitative only in relation to other attitudes in a representation of attitudes. Some attitudes are intrinsically quantitative and not just quantitative in a representation of comparative relations among attitudes.

Intrinsic quantities, that is, *magnitudes*, are not comparative relations among objects. An object's length is a property of the object and not just a representation of its comparative relations with other objects. An agent may have direct access to an object's length without inferring it from the object's relations to other objects. An agent may see a stick's length.[9]

---

[9] Although nonquantitative magnitudes may exist, I use the term magnitude only for quantitative magnitudes. A magnitude is an intrinsic property of an object and, being quantitative, it is apt for

Introspection may provide access to a quantitative propositional attitude, as vision provides access to a table's length. An agent may directly know her certainty that Socrates died in 399 BC or did not. She may directly know ½ is her degree of belief that heads will turn up on a toss of a fair coin. Similarly, an agent may have direct access to quantitative comparisons of her attitudes to propositions. She may directly know that she believes that a red card will appear on a random draw from a standard deck of cards twice as strongly as she believes that a diamond will appear, besides directly knowing the strength of her doxastic attitude that a red card will appear. Introspection gives humans some access to their quantitative propositional attitudes, even if not certainty of their attitudes, and gives ideal agents reliable access to their quantitative propositional attitudes.

A propositional attitude, I assume, may have a magnitude that does not reduce to its relations to other propositional attitudes. Its having such a magnitude is its having an apt numerical representation that shows the relation of its magnitude to other perhaps unrealized magnitudes of the propositional attitude. The numerical representation need not show the relation of the attitude to other attitudes an agent has formed. An agent who knows that the physical probability of drawing a red ball from an urn is ½ may have a degree of belief of ½ that red will be drawn without having degrees of belief concerning other independent propositions. The degree of belief is an apt representation of the magnitude of the attitude in relation to other magnitudes such as certainty, even if the agent is not certain of any proposition. An attitude having such a magnitude is intrinsically quantitative.

Although a quantitative attitude may exist without being accessible, such attitudes are sometimes accessible through introspection. Even without the direct access introspection brings, the attitude is inferable from brain scans it causes, behavior it causes, the causes of the attitude, and similar data. Introspection and other forms of access are evidence of the existence of quantitative attitudes, taken as theoretical entities revealed by observations although not defined by observations.

One may represent empirical relations with numerical relations, and the empirical relations represented need not be observational. One may infer the

---

representation on a numerical scale showing its relation to other magnitudes of the same type. Although its representation on the scale uses its relation to the scale's parameters, such as a unit for a ratio scale, its aptness for representation on a numerical scale is an intrinsic property. A stick's length has this aptness, whereas the property of being mud does not. If the only object existing were a stick (and its parts), the stick could double in length. Its doubling in length is a relation of its current length to its prior length, a relation of properties and not a relation of objects. The two properties exist at the same time even if no two objects existing at the same time have the two properties.

empirical relations that one represents numerically. So, even if ratios of degrees of belief were not introspectable but only inferable, their numerical representation may proceed. An observer may infer a ratio to represent. Among the causes of a ratio are evidence concerning its components and among the effects are choices involving its components. The causes and effects of the ratio are grounds for inferring it. An observer may infer a ratio, such as an agent's believing in heads half as strongly as believing in heads or tails, from the agent's betting behavior or from the agent's statistical data.

This subsection and the previous two subsections sketch a realist account of precise probabilities and utilities, which Weirich (2001) presents more thoroughly. The next subsection presents a realist account of imprecise probabilities and utilities.

2.5 Representing Imprecision

Principles of rational choice apply using a representation of an agent's doxastic and conative state and not using the state itself. Although the state itself is metaphysically determinate, its representation in a decision problem may be indeterminate.[10] Hence a choice's compliance with a principle of rational choice may be indeterminate. For example, if the options in the decision problem have indeterminate utilities, then whether a choice maximizes utility may also be indeterminate. As Subsection 2.1 notes, I investigate cases in which an agent's doxastic and conative state has a determinate representation adequate for rational choice, even if the representation imprecisely specifies probabilities and utilities.

The purpose of a representation settles whether it is adequate. In a decision problem, I assume a representation of an agent's doxastic and conative state, including the doxastic and conative attitudes it contains, that is adequate for the purpose of evaluating choices for rationality and more generally evaluating options for rationality (as some options are not realized). A doxastic attitude to a proposition may have a representation as a belief and also a representation as a degree of belief. Representing the doxastic attitude as a belief is inadequate for evaluation of a choice because no general principle of rationality for choice uses beliefs.

A representation of an agent's doxastic and conative state need not represent all features of the state (just as a map of a city need not represent all features of the city). Decision principles need a representation of the state adequate for evaluating the options in a decision problem. To assess the adequacy of a representation, I adopt a general principle of rational choice for an agent

---

[10] I assume metaphysical determinacy and so the absence the phenomenon called ontic vagueness.

facing a decision problem in the model presented – a principle supported by judgments about rational choice – and then select from among various representations of the agent's doxastic and conative state a representation that furnishes the input for the decision principle.

A precise probability inadequately represents some doxastic states because the precise probability leads via maximization of expected utility to a single choice, whereas, given the agent's imprecise probabilities and utilities, typically many options are permissible – even in the absence of indifference between any two options in the decision problem. It may be permissible to proceed with a picnic, or to cancel the picnic, given an imprecise probability of rain, an imprecise utility of a picnic without rain, and an imprecise utility of a picnic with rain – even if the agent is not indifferent between proceeding and cancelling. Given her imprecise probabilities and utilities, she need not be indifferent between the two acts, or even have the same conative attitude to the two acts, for each to be permissible.

In the decision model, a set of pairs of a probability function and a utility function for the possible outcomes of the options in a decision problem provides a representation of an agent's doxastic and conative state adequate for a rational resolution of the decision problem. I call the set *the representative set*. The representative set settles the options that are permissible given the agent's doxastic and conative state concerning the possible outcomes of options. An option is permissible if and only if it maximizes expected utility according to a pair of a probability function and a utility function in the representative set. This is *the permissive principle of choice* that Section 7 advances. A version of it appears in Good (1952: 114).[11]

Suppose that for an agent with a choice between two acts, such as proceeding with a picnic or cancelling the picnic, the two acts are incomparable because of imprecise probabilities and utilities for possible outcomes. The representative set for the agent's doxastic and conative state then has a pair of a probability function and a utility function according to which the first act maximizes expected utility and also has a pair of probability function and a utility function according to which the second act maximizes expected utility. Applying the permissive principle of choice, both acts are rational.

This section uses the permissive principle of choice for decision problems with imprecise probabilities and utilities to argue that the representative set is an adequate representation of an agent's doxastic and conative state. Section 7 supports the permissive principle of choice on grounds independent of the set's being an adequate representation.

---

[11] In Levi's (1980) terminology, an option that the permissive principle sanctions is *e-admissible*.

A representation that is richer than a set of pairs of a probability function and a utility function may also be adequate. A representation that along with the set of pairs includes a weight for each pair, such as a probability of the pair, is also adequate.[12] However, as Section 7 argues, the weights are not necessary for identifying rational choices in a decision problem. The probability-utility pairs of the representative set form a *minimal* adequate representation. They represent all and only the features of an agent's doxastic and conative state that a general principle of rational choice uses for decision problems in the decision model. Fewer features may suffice in special cases, for example cases with a dominant option, but they do not suffice in all cases and so are not adequate generally. Assuming the permissive principle of choice, the representative set forms a minimal adequate representation of an agent's doxastic and conative state for all decision problems arising in the decision model.

Besides a representation of an agent's doxastic and conative state using a set of pairs of a probability function and a utility function, theorists advance other rival representations. I review a few candidates for being a minimal adequate representation in the general case. The candidates offer a representation of an agent's nonquantitative doxastic state assuming that the agent's utility assignment is precise. In this case, the representative set has pairs of a probability function together with a constant utility function, and the set of probability functions in the pairs represents the agent's doxastic state. The candidates propose an alternative to this set as a representation of the agent's doxastic state.

A propositional attitude may have a quantitative representation without being quantitative. Numbers may represent only comparisons of attitudes. If an agent more strongly believes that it will rain than that it will not rain, then assigning 0.6 as the agent's probability that it will rain and 0.4 as the agent's probability that it will not rain represents the comparison even if the agent does not have quantitative doxastic attitudes to the two propositions.

Weirich (2004: sec. 4.2) discusses a best point-representation of nonquantitative doxastic attitudes. According to this representation, a precise probability is a best representation, using just a single probability assignment, of a nonquantitative doxastic state. This representation is not adequate because using it together with an expected-utility evaluation of options rules out choices that rationality permits; in a decision problem without indifference between any options, it deems a single option rational although in some cases multiple options

---

[12] Lassiter (2020) argues for a representation of a doxastic state that uses probabilities of probabilities, claiming that a representation that uses sets of probability assignments is not adequate for an account of learning from experience. Even granting this objection, which Subsection 3.1 rebuts, a set of probability assignments may nonetheless be adequate for an account of rational choice.

are rational. For example, if a best point-representation of the probability of rain is 0.6, then if using that probability cancelling a picnic has higher expected utility than holding the picnic, cancelling is the unique rational choice according to the representation, although holding the picnic is also rational if it maximizes expected utility according to another probability assignment in the set of assignments that represents the agent's nonquantitative doxastic attitude to rain. A best point-representation glosses over features of doxastic attitudes that affect rational choice. It treats a choice resting on nonquantitative doxastic attitudes to possible outcomes the same as a choice resting on quantitative doxastic attitudes to possible outcomes, although, I assume, rational choice attends to the difference between nonquantitative and quantitative doxastic attitudes.

Intervals are not good representations of nonquantitative doxastic attitudes. Consider a representation of the attitude to p with the interval $[0.6, 0.9]$, the attitude to $\sim p$ with the interval $[0.1, 0.4]$, and the attitude to $(p \vee \sim p)$ with the interval $[1, 1]$. These intervals do not display some significant relations between the attitudes to p and to $\sim p$. For example, they do not show that as one attitude strengthens the other weakens, so that certainty is the constant attitude to $(p \vee \sim p)$. Also, consider the interval $[0.2, 0.4]$ for p and $[0.3, 0.5]$ for q. They do not indicate whether p and q are doxastically compared. It may be that in the representative set every assignment of probabilities assigns a lower probability to p than to q, so that the doxastic attitude to p is weaker than the doxastic attitude to q. Or, it may be that some assignments put p lower than q while others put p higher than q so that no doxastic comparison holds between p and q. Furthermore, adding an interval $[0.06, 0.2]$ for $(p \ \& \ q)$ does not show whether p and q are independent in the sense that the probability of $(p \ \& \ q)$ equals the probability of p times the probability of q. The numbers in the interval for $(p \ \& \ q)$ may, but need not, be products of the numbers in the intervals for p and for q, respectively. Intervals are not as expressive as a set of probability assignments.

An agent's doxastic attitude to a proposition is just one element of the agent's doxastic state, which includes the agent's doxastic attitudes to all propositions in the agent's doxastic domain, even propositions without a precise probability assignment. Because an adequate representation of an agent's doxastic attitudes must represent their relations, I assume a representative set for an agent's doxastic state and use it to construct representations of an agent's doxastic attitudes. An interval representation of a doxastic attitude to a proposition displays only the interval of values that the probability assignments in the representative set attribute to the proposition, and this interval has no role in a general decision principle.

An agent's views about relations among events may affect a rational choice. So, representing a nonquantitative doxastic attitude with an interval is not adequate for rational choice; representing the doxastic attitude to each possible outcome of an option with an interval may lead to an irrational choice. Suppose that an agent's interval for rain is [0.3, 0.7] and is also [0.3, 0.7] for rain's absence. Imagine that the agent must decide whether (1) to accept a gamble that pays $10 if it rains and $10 if it does not rain or (2) to accept $8. Using a variation of the permissive principle that allows her to decide using any probabilities in the intervals she assigns to rain and to rain's absence, she may accept $8 because using 0.3 for rain and 0.3 for rain's absence makes the expected payoff of the gamble equal $6. However, she should accept the gamble because it guarantees $10.

In contrast with intervals, a set of probability assignments adequately represents an agent's doxastic state. Combined with the permissive principle of choice, the set correctly classifies the options in a decision problem according to rationality. An evaluation of an option using the agent's representative set compares the expected utilities of options according to each probability assignment in the set. In the example, no assignment makes the probability of rain 0.3 and the probability of rain's absence 0.3. Under each assignment, accepting the gamble has an expected payoff of $10 and so has greater expected utility than accepting $8. Following the permissive principle, an evaluation of options using the representative set makes accepting the gamble the rational choice.

2.6 Measurement

Rational choices need not reveal imprecise probabilities and utilities. For example, suppose that an agent has imprecise probabilities and utilities for the possible outcomes of holding a picnic as planned or cancelling it because of the threat of rain. If the agent cancels, this choice does not show that the expected utility of cancelling is greater than the expected utility of holding the picnic, for the expected-utility comparison assumes precise probabilities and utilities. If a rational ideal agent's choices have a unique representation according to which they maximize expected utility (given a unit for utility), the probability and utility functions of the representation are precise. Hence, if the agent's doxastic and conative attitudes have imprecise representations, the precise functions do not represent them.[13]

---

[13] Seidenfeld, Schervish, and Kadane (2010) take a choice to be admissible if it maximizes expected utility according to an admissible pair of a probability function and a utility function. They show that a choice function satisfies four axioms of coherence only if it can be represented as a coherent choice function according to a set of probability-utility function pairs that meets certain conditions and that may represent imprecise probabilities and utilities. This result does not establish that imprecise probabilities and utilities may be measured using coherent choices, because it does not establish that the representation is unique given a unit for utility. Seidenfeld, Schervish, and Kadane (2012) investigate elicitation of imprecise probabilities under some

Taking the construction of probability and utility functions using choices as a means of measuring rather than defining probabilities and utilities, the uniqueness of the probability function and (given a unit) the utility function representing choices assumes that choices issue from precise probabilities and utilities that the representation reveals. When this assumption is inaccurate, the representation misrepresents the doxastic and conative attitudes that yield the choices. The representation assumes that an option realized in a decision problem is the unique rational resolution of the problem (given the decision model's stipulation that different options have different utilities). However, in a decision problem, if possible outcomes of options have imprecise probabilities and utilities, so that options have imprecise expected utilities, multiple options may be rational, not just the option adopted. Measurement of imprecise probabilities and utilities faces challenges.

2.7 Summary

This section takes probabilities and utilities to represent propositional attitudes in quantitative cases. Then it extends the view, taking a set of pairs of a probability function and a utility function to represent an agent's doxastic and conative state in nonquantitative cases. It assumes that, in the decision model constructed, rational degrees of belief are probabilities and that rational degrees of desire are utilities. Sections 4 and 5 support this assumption. To assess the adequacy of a representation of an agent's nonquantitative doxastic and conative state, the section assumes the permissive principle of choice that Section 7 advances.

3 Rational Imprecision

Rationality requires that the doxastic and conative attitudes of ideal agents be such that their representations for decision problems have certain mathematical properties. This section maintains that rationality permits, and may require, their representations to be imprecise. Sections 4 and 5 argue that degrees of belief, if rational, obey the laws of probability and that degrees of desire, if rational, obey the expected-utility principle so that they qualify, respectively, as probabilities and utilities. Section 6 extends these requirements to pairs of

---

assumptions about the scoring rule for forecasts and about an agent's decision rule. Under their assumptions, an agent expects her announcements of imprecise probabilities to receive a good score if and only if her announcements reveal her imprecise probabilities. This elicitation, or measurement, of imprecise probabilities uses announcements rather than choices in general. Also, the lexicographic scoring rule they assume and the decision rules they assume are controversial; the decisions rules conflict with the permissive principle that Section 7 supports.

a degree of belief assignment and a degree of desire assignment in a set of pairs that represents nonquantitative attitudes.

3.1 Sources and Consequences of Rational Imprecision

I take rationality not as a mental capacity but as a source of standards for mental states and for acts. For example, it imposes standards for doxastic attitudes, such as consistency for the beliefs of ideal agents in ideal circumstances for forming beliefs. Irrationality entails being blameworthy and so is sensitive to an agent's abilities and circumstances. Because departures from rationality's standards are blameworthy, the standards are less demanding for humans forming doxastic and conative attitudes in distracting circumstances than they are for ideal agents in ideal circumstances for forming these attitudes. The standards of rationality I advance for doxastic and conative attitudes understand rationality in the ordinary normative sense, not in a technical sense that makes the standards hold by definition, so I justify the standards.

The standards for degrees of belief and degrees of desire assume, as Section 2 explains, that degrees of belief and degrees of desire represent quantitative propositional attitudes and are not constructed from comparative doxastic relations, preferences, or choices. In cases of imprecision, a set of degree of belief assignments represents an agent's doxastic state, which includes all the agent's doxastic attitudes. The state's rationality does not imply that every assignment in the set is warranted by the agent's evidence; an assignment taken in isolation may exhibit spurious precision. The rationality of the doxastic state shows in features of the whole set. Similarly, in cases of imprecision a set of degree of desire assignments represents a conative state, and the state's rationality shows in features of the whole set rather than in each assignment. In fact, if evidence and experience are sparse, and rationality requires imprecision, then each precise assignment of degrees of belief and of degrees of desire, taken by itself, represents an irrational doxastic and conative state. Although the precise assignments, by complying with the laws of probability and utility, meet some requirements of rationality, their precision is unjustified.

Lack of information is a reason for imprecise degrees of belief. Lack of experience is a reason for imprecise degrees of desire, and lack of information is an additional reason for imprecise degrees of desire that depend on information. Consequently, in a decision problem, lack of experience and lack of information together are reasons for imprecise degree of belief assignments to possible outcomes of options and imprecise degree of desire assignments to options and their possible outcomes.

Rationality permits a precise degree of belief only given sufficient evidence to warrant the precision and permits a precise degree of desire only given sufficient experience and information to warrant the precision. Rationality prohibits a precise degree of desire that a proposition hold without sufficient understanding of the proposition's consequences. If because of insufficient experience and information, a rational ideal agent's understanding of the consequences of a pair of propositions is insufficient for comparing them, they are incomparable for the agent and so do not both have precise utilities.

For degrees of belief, I assume a doxastic domain in which, relative to no evidence, a proposition that is possible is an epistemic possibility and has a positive probability. I also assume that evidence settles a rational doxastic attitude to a proposition, although I do not have a general account of this. As an illustration, the Principal Principle (a type of principle of direct inference) requires assigning to an event with a known chance a degree of belief equal to its chance, assuming no additional relevant evidence concerning the event.[14] This principle describes how in special cases evidence settles a degree of belief.

As an idealization, I assume that an agent learns by acquiring certainties. She suffers no memory losses and never loses evidence; moreover, she knows that this is so. These assumptions remove objections to taking standard conditionalization as a requirement of rationality for updating assignments of degrees of belief as new evidence arrives.[15] To formulate the requirement, let P stand for probability, h stand for a hypothesis, e stand for the total new evidence, and $P(h|e)$ stand for the probability of h given e, taken as $P(h \& e)/P(e)$. Thus, according to the requirement, if just before the new evidence e arrives $P(h|e) = x$, then immediately after the new evidence e arrives $P(h) = x$.[16]

The primary norm for degrees of belief is fit with total evidence. I take the norm of conditionalization to derive from the primary norm's requiring degrees of belief that at each moment fit the evidence at the moment. I assume that conditionalization for degrees of belief follows from rationally responding to each evidential state in a sequence of evidential states. The norm that has evidence settle degrees of belief leaves no room for an independent diachronic norm.[17]

---

[14] A general version of the principle relaxes the assumption about the absence of additional relevant information and requires only the absence of *inadmissible evidence*, such as evidence about the outcome of the chance for the event.

[15] The assumption that evidence arrives in the form of certainties also puts aside the case for a generalization to Jeffrey conditionalization.

[16] This version of conditionalization makes explicit that conditionalization uses as a condition a proposition that expresses total new evidence. Conditionalization is not a satisfactory updating procedure for me if, for example, the conditioning proposition is that I have acquired more evidence than this very proposition.

[17] The priority of synchronic standards of rationality emerges from time-slice rationality as Hedden (2015) describes it.

Subjectivists deny that standards of rationality settle a unique doxastic attitude to a proposition given a body of evidence. I agree that standards do not always settle a unique precise degree of belief but assume that they always settle a unique doxastic state that a set of degree of belief assignments represents. The case for denying this assumption appears to be just a case for tolerance of divergent doxastic states in the absence of consensus on how standards of reasoning, given a body of evidence, settle a unique doxastic state. Tolerance is in order for the doxastic states of humans, given their ignorance of rationality's standards for fit with evidence. It is also in order for a human theorist's treatment of the doxastic states of ideal agents. Even if the ideal agents know rationality's standards, a human theorist, who does not know the standards, is not in a position to assert the effect of the standards on the doxastic state of an ideal agent. Nonetheless, the case for tolerance does not undermine the assumption that rationality settles a unique doxastic state fitting a body of evidence.[18]

Reasons that do not motivate ideal agents motivate humans to be cautious with assignments of extreme degrees of belief. A real person, for safety, may not assign maximum degree of belief to any proposition, not even to a tautology, fearing a mistake in identifying tautologies. However, an ideal agent knows she makes no mistakes identifying tautologies and without fearing a mistake assigns maximum degree of belief to every tautology she considers. Although rational people fail to assign maximum degree of belief to all tautologies, a rational ideal agent assigns maximum degree of belief to every a priori truth, including every tautology, to which she assigns a degree of belief.

An ideal agent, assuming she acquires evidence in the form of certainties, does not comply with the principle of regularity, which prohibits assigning an extreme degree of belief to any a posteriori proposition. Although humans have reason to comply with the principle, because of their cognitive limits and the fallibility of their information, they also have a countervailing reason, for the sake of simplicity, sometimes to assign an extreme degree of belief to an a posteriori proposition; in some cases an extreme assignment conveniently puts aside a negligible difference from 0 or 1. After assigning an extreme degree of belief to a proposition, if a person acquires relevant evidence, she may discard her earlier simplification to obtain a better fit with her total evidence. After acquiring relevant evidence, she may start afresh with assignments of degrees of belief; the past does not bind her, via conditionalization, to continuing to assign an extreme degree of belief to the proposition

---

[18] Jackson and Turnbull (forthcoming) review arguments for and against uniqueness and alternatives to uniqueness. Greco and Hedden (2016) support uniqueness. Castro and Hart (2019) present a dilemma for supporters of uniqueness that my idealizations about agents resolve.

In contrast, a rational ideal agent does not assign a degree of belief 0 or 1 to a proposition unless the evidence is conclusive. Although it usually is not, and degrees of belief for a posteriori propositions are typically between 0 and 1, when evidence justifies an extreme degree of belief, conditionalization retains the extreme assignment as new evidence arrives, and this does not entail any assignments of degrees of belief that do not fit the agent's evidence at a time. An ideal agent has no reason to make a fresh start with assignments of degrees of belief. If an ideal agent rationally gives an a posteriori truth, such as the agent's seeing red, a degree of belief 1, the agent rationally maintains her certainty as new evidence arrives. Rational ideal agents, following conditionalization, persist with assignments of degree of belief 0 and degree of belief 1 once given.[19]

Consider an agent with a doxastic state that a set of degree of belief assignments represents. Suppose that the agent is limited rather than ideal. Acquiring new relevant evidence may give the agent a reason to drop some degree of belief assignment in the set, perhaps an assignment that attributes an extreme degree of belief to an a posteriori proposition. Dropping an assignment is a way of starting afresh. Rationality may grant to humans this liberality in updating. However, ideal agents need not correct cognitive shortcuts when they update. For ideal agents, rationality requires updating a doxastic state after acquiring new relevant evidence by conditionalizing, on the total new evidence, every degree of belief assignment in the set of assignments representing the doxastic state.

If an ideal agent's doxastic state has a representation by a set of degree of belief assignments such that a proposition has degree of belief 0 in one assignment and degree of belief 1 in another assignment, then assuming that updating works by conditionalizing each degree of belief assignment in the set, the representative set continues indefinitely to have an assignment giving the proposition degree of belief 0 and another assignment giving the proposition degree of belief 1, no matter what evidence arrives. This result may seem to ignore new relevant evidence that justifies raising an assignment of 0 or lowering an assignment of 1. However, a rational ideal agent attributes degree of belief 0 and degree of belief 1 only when justifiably certain that no such new relevant evidence will arrive.

Imagine a toss of a coin of unknown bias and let h be the hypothesis that it lands heads. Suppose that an agent has a set of degree of belief assignments that collectively give the hypothesis h every number in the interval [0, 1]. Imagine that the agent acquires just the evidence e that on the previous ten tosses the coin came up heads five times. After conditionalization on the total new evidence e,

---

[19] Weisberg (2009) rejects this rigidity, however my idealizations put aside his objections to it.

for every number in the interval, some degree of belief assignment in the agent's set gives h a degree of belief equal to that number. It may seem that conditionalization prevents learning from evidence e that bears on the hypothesis h.[20] However, under the idealizations about agents, this case does not arise because an agent considers all epistemically possible bodies of future evidence and, if any warrants degree of belief assignments different from 0 or 1, does not currently assign 0 or 1 to the hypothesis. If a hypothesis and its negation are both epistemically possible, then a representative degree of belief assignment to the hypothesis falls into a closed interval with non-extreme endpoints. Given no evidence, every future body of evidence in the agent's doxastic domain that is possible is epistemically possible and does not receive degree of belief 0 in any assignment in the set representing the agent's doxastic state. No big surprises occur because no possible future event is epistemically impossible.[21] Given the idealizations, conditionalization is an appropriate method of updating both a degree of belief assignment and a set of degree of belief assignments. Imprecision does not block learning from experience.

3.2 Dilation

Some theorists argue that having imprecise probabilities and utilities is irrational. Section 8 defends imprecise probabilities and utilities against the charge that they lead to incoherent decisions. This subsection defends doxastic imprecision, assuming that a set of probability functions represents it, against an objection concerning the effect of new information.

The objection rejects imprecise probabilities because, in contrast with precise probabilities, they lead to cases of *dilation*, that is, cases in which new information makes an agent's probability assignment less, rather than more, precise. It regards dilation as irrational and claims that a doxastic state is irrational if its representation by a set of probability functions leads to dilation.

New evidence may increase probabilistic precision about some propositions but decrease probabilistic precision about other propositions. For some propositions, new evidence can undo the case for a precise probability assignment and support making their probability assignments imprecise. There are many unsurprising cases of this sort. Suppose that an agent assigns probability 90 percent to

---

[20] Moving from the closed interval [0, 1] to the open interval (0, 1) does not make possible learning from experience because using new evidence to conditionalize the degree of belief assignments that generate the elements of the open interval regenerates the open interval, as Castro and Hart (2019) note.

[21] The idealizations rule out Gareth Evans (1979) cases, such as the possibility of Julius not being the inventor of the zip, although for the agent it is epistemically impossible that Julius is not the inventor of the zip because, by dubbing, he is the inventor of the zip.

rain using her weather radar and then learns that her weather radar is broken. Not having other clues about rain, her probability assignment for rain may become imprecise. Not only is dilation rational but also assigning a precise probability to a hypothesis does not prevent dilation given new evidence, if the probability of the hypothesis conditional on the new evidence is imprecise.

Dilation may seem irrational because in many cases new relevant evidence increases the precision of a doxastic attitude to a hypothesis. However, sometimes new evidence muddies the case for a hypothesis so that its probability assignment becomes more imprecise. Titelbaum (forthcoming: sec. 14.2.2) claims that in cases starting with specific information about an event that grounds a precise probability assignment to the event, dilation is the appropriate epistemic response to new evidence that is unspecific about the event.

The claim that dilation is irrational may restrict itself to cases in which dilation occurs for a proposition no matter which element of a partition one learns and the element learned has no direct bearing on the proposition. White's (2010) objection to imprecise probabilities uses cases of this sort.

Joyce (2010) presents a version of White's (2010) coin case to illustrate allegedly problematic dilation. Joyce's presentation uses C, a set of credence functions, to represent an agent's credal state. Each credence function c is a probability function. In my terminology, a credal state is a doxastic state and a credence function is an assignment of degrees of belief.

> **Coin Game**. I have a *fair* coin with heads on one side and tails on the other and a coin of *entirely unknown bias* that is black on one side and gray on the other. Since you know the heads/tails coin is fair, each c in your credal state will satisfy $c(H) = \frac{1}{2}$, where H says that the coin comes up heads on its next toss. Since you know nothing about the black/gray coin, the imprecise interpretation says that there will be a c in your credal state with $c(B) = x$ for every $0 < x < 1$. I will toss the coins, independently, and observe the results without showing them to you. I will then tell you something about how the outcomes are correlated by reporting either "$H \equiv B$" (if I see heads and black or tails and gray) or "$H \equiv \sim B$" (if I see heads and gray or tails and black). Since the heads/tails coin is fair and tosses are independent, you anticipate these reports with equal probability: $c(H \equiv B) = c(H \equiv \sim B) = \frac{1}{2}$ for all $c \in C$ (Joyce 2010: 296–297).

Suppose that you learn that $H \equiv B$. Because your credence functions follow the laws of probability,

For every $c \in C$,

° the credence that c assigns to B upon learning $H \equiv B$ is identical to the *prior* credence that c assigns to B, $c(B|H \equiv B) = c(B)$;

° the credence that c assigns to H upon learning $H \equiv B$ is identical to the credence c assigns to B upon learning $H \equiv B$, $c(H|H \equiv B) = c(B|H \equiv B)$.

This is surprising. Learning that H and B are perfectly correlated provides you with *no relevant evidence* about B but it forces you to conform your beliefs about the outcome of the heads/tails toss (about which you know a lot) to your prior beliefs about B (about which you know nothing) (Joyce 2010: 297–298).

White holds that such dilation of the credence that H is irrational. However, Joyce disagrees. He observes that each biconditional you may learn is relevant to H according to the credence functions in the set representing your doxastic state. He therefore defends the rationality of the dilation with these two points.

○ Both $H \equiv B$ [and] $\sim H \equiv B$ are highly evidentially relevant to H even when you are entirely ignorant about B. It only seems otherwise because their evidential impact on H's probability is indeterminate, both in magnitude and in direction.

○ Given this, it would be wrong to retain a credence of ½ for H upon learning $H \equiv B$ or $\sim H \equiv B$. Instead, you should have an indeterminate credence for H that reflects the indeterminate evidential import of your new evidence (Joyce 2010: 299).

Hart and Titelbaum (2015) also argue that the dilation is not irrational. They take dilation to be surprising in White's coin game because, in general, learning a biconditional has surprising effects, even in cases with precise probabilities. Also, even when the information acquired is not a biconditional, Pedersen and Wheeler (2014) show, given a nuanced account of relevance, that allegedly problematic dilation is not irrational if it arises in a properly constructed probability model. Because of such defenses of dilation, the phenomenon does not discredit imprecise doxastic attitudes.

4 Probabilism

Decision principles address a decision problem using a representation of an agent's doxastic and conative state regarding the options in the decision problem. An adequate representation of a decision problem uses a set of pairs of a degree of belief assignment to the options' possible outcomes and a degree of desire assignment to the options' possible outcomes and to the options themselves. Decision principles assume that rationality's standards for an agent's doxastic and conative state impose laws of probability on degrees of belief and laws of utility on degrees of desire.

If in a decision problem an agent's doxastic and conative attitudes to the options' possible outcomes are quantitative, a representation of the agent's doxastic and conative state has a single pair of an assignment of degrees of belief and an assignment of degrees of desire. This section and Section 5 treat this special case. They show that in the decision model constructed rationality requires the agent's state to be such that the degree of belief assignment is a probability function and the degree of desire assignment is a utility function. This section treats rationality's standards for the degree of belief assignment, and Section 5 treats rationality's standards for the degree of desire assignment. Section 6 argues that in general rationality requires an agent's doxastic and conative state to be such that a set of pairs of a degree of belief assignment and a degree of desire assignment representing the state have assignments of degrees of belief that are probability functions and assignments of degrees of desire that are utility functions. Argumentation supporting norms for precise attitudes lays the groundwork for argumentation supporting norms for impre-cise attitudes.

4.1 Probabilism Defined

Probabilism is the view that rational degrees of belief for the propositions in a doxastic domain comply with the axioms of probability and hence with the laws of probability that follow from the axioms. It addresses degrees of belief at a single time and so given the same evidence. In some cases, an agent does not assign degrees of belief to all the propositions in a doxastic domain. However, the version of probabilism I consider applies only to an agent who, having sufficient evidence, justifiably assigns degrees of belief to all the propositions of a finite doxastic domain.[22] It claims that these degrees of belief, if the agent is ideal and rational, conform to the probability axioms.

   I assume that probability P attaches to propositions and use the propositional variables p and q to state the standard Kolmogorov axioms of probability:

Non-negativity: $P(p) \geq 0$.
Normality: $P(p) = 1$, if p is a tautology.
Additivity: $P(p \vee q) = P(p) + P(q)$, if p and q are logically exclusive.[23]

The axioms assume that P is a function from the propositions in a doxastic domain to real numbers and hence that $P(p)$ has a unique value. I call the assumption that $P(p)$ is a unique real number *singularity*.

   The argument for probabilism assumes that an agent is cognitively ideal and conditions are ideal for having degrees of belief that comply with singularity and the axioms of probability. Ideal conditions rule out cases in which an agent

---

[22] Restricting the assignment to the propositions in a finite doxastic domain puts aside the issue of countable additivity.

[23] More precisely, additivity assumes that p and q are logically exclusive according to propositional logic.

may save the world by violating an axiom of probability. They remove obstacles to complying with the laws. Compliance violates no norm of rationality concerning practical consequences.

The axiom of additivity, taken as a requirement of rationality, may have either wide or narrow scope. According to a wide-scope interpretation, the axiom requires that if a disjunction has incompatible disjuncts, then given degrees of belief for the disjunction and its disjuncts, the degree of belief that the disjunction holds equal the sum of the degrees of belief that the disjuncts hold. According to a narrow-scope interpretation, the axiom requires, given degrees of belief for the disjuncts, that the degree of belief for the disjunction equal their sum. Under the wide-scope interpretation, the requirement governs the relation between three degrees of belief, whereas under the narrow-scope interpretation it governs only the degree of belief that the disjunction holds.

Using O to express an obligation of rationality that a proposition hold and assuming that p and q are logically exclusive, according to the wide-scope interpretation, $O(P(p \lor q) = P(p) + P(q))$, whereas, according to the narrow-scope interpretation, if $P(p) + P(q) = x$, then $O(P(p \lor q) = x)$. Thus, given the narrow-scope interpretation, if for an agent $P(p) = P(q) = \frac{1}{6}$, then for the agent $P(p \lor q)$ must equal $\frac{1}{3}$. However, given the wide-scope interpretation, this obligation does not exist. The agent can satisfy the wide-scope norm despite making $P(p \lor q) = \frac{1}{2}$, if she changes her probability assignments to the disjuncts so that $P(p) = P(q) = \frac{1}{4}$.

The narrow-scope interpretation yields the stronger norm because, given a probability assignment to p and to q, it requires a particular additive assignment to $(p \lor q)$ and not just any additive assignments to p, q, and $(p \lor q)$. The narrow-scope norm imposes an obligation that may not be met just by meeting the obligation that the wide-scope norm imposes. Meeting the narrow-scope norm entails meeting the wide-scope norm but the converse entailment does not hold. In the example, the narrow-scope norm would not be met if the agent were to adopt the assignments $P(p) = P(q) = \frac{1}{4}$ and $P(p \lor q) = \frac{1}{2}$ because, in view of her actual assignments to p and to q, she has an obligation to assign $\frac{1}{3}$ to $P(p \lor q)$.

I take probabilism to advance the wide-scope interpretation of additivity rather than the stronger, narrow-scope interpretation. However, the idealizations adopted make the two interpretations equivalent. If (1) an agent is ideal and rational except possibly in the matter of compliance with the norm of additivity, (2) has formed rational degrees of belief for the disjuncts but not yet formed a degree of belief for the disjunction, (3) her degrees of belief for the disjuncts are rational, and (4) forming a degree of belief for the disjunction equal to their sum is not contrary to any norm of rationality; then if rationality requires her to comply with the wide-scope norm, it also requires her to comply with the stronger, narrow-scope norm. The agent may not adopt a degree of belief for the disjunction that violates the narrow-scope norm and then change her degrees of belief for the disjuncts to comply with the wide-scope norm because her evidence settles rational degrees of belief for the disjuncts.

Intuition supports probabilism for ideal agents. An intuition that rationality requires compliance with an axiom is an inference of the requirement from a full understanding of a proposition expressing the requirement. The inference proceeds not from other propositions but from an understanding of the proposition's parts and structure. To illustrate such an inference, consider the proposition that $1 = 1$. Although the proposition may be inferred from the generalization that everything equals itself, it may also be inferred from an understanding of 1 and identity.

Although probabilism has the support of intuition, many theorists also seek an argument that infers probabilism from other propositions. Subsection 4.5 presents such an argument. To set the stage for the argument, I consider how best to argue for probabilism.

4.2 Types of Argument for Probabilism

A good argument that rational degrees of belief comply with the axioms of probability attends to the interpretation of degrees of belief. A definition of degrees of belief as the values of a probability function constructed to represent preferences among gambles stipulates that degrees of belief comply with the axioms of probability. Given the definition, an argument for probabilism can simply appeal to the definition. Because I take degrees of belief as representations of propositional attitudes, their compliance with the axioms of probability is not a matter of definition but a normative requirement. Given my formulation of probabilism, an argument for it must be normative.

Degrees of belief direct choices and if they have irrational features may produce irrational choices even when rationally used. An argument for probabilism may point out the bad consequences for choices of degrees of belief in violation of the axioms of probability. The sure-loss argument for probabilism, which Titelbaum (forthcoming: chap. 9) reviews, shows that degrees of belief that violate the probability axioms may lead to a system of bets that guarantees a net loss and that complying with the axioms prevents such a system of bets. The argument is pragmatic and instrumental. It presents a financial risk of violating the axioms. Even if the argument succeeds, a theorist may seek a different, epistemic argument for probabilism because the view advances epistemic norms for degrees of belief. Probabilism, as I present it, takes degrees of belief to represent doxastic attitudes and so invites an epistemic argument for their compliance with the probability axioms.

An epistemic argument for probabilism assumes that rational degrees of belief have their role in choice because of their compliance with the probability axioms and do not comply with the probability axioms because of their role in choice. One epistemic argument shows that violations of the axioms yield inconsistent conclusions about the fairness of bets, as Howson and Urbach (2006: chap. 3) and Christensen (2004: chap. 5) observe. For example, a violation of normality leads to the judgment that a fair bet may wager against the truth of a tautology. Such an argument, although not pragmatic, is instrumental. It shows how compliance with the axioms serves the epistemic goal of consistent beliefs.

Joyce's (1998) argument for probabilism is also epistemic. It notes that degrees of belief have the epistemic goal of being 1 for truths and 0 for falsehoods and are more accurate the closer they are to the goal. It shows that, if an assignment of degrees of belief violates the probability axioms, then another assignment is sure to be more accurate (given an acceptable measure of accuracy). This provides an epistemic reason to comply with the axioms, given the existence of compliant assignments not sure to be less accurate than rivals. The argument shows that compliance with the axioms is a means of furthering the epistemic goal of graded accuracy. This epistemic argument is instrumental. It shows that compliance with the probability axioms is a means of promoting graded accuracy.

Accuracy-first epistemology, and its elaboration in epistemic utility theory, takes accuracy to be the primary epistemic goal and all other epistemic goals to derive from the goal of accuracy.[24] However, justification by evidence is a traditional, independent epistemic goal for both belief and degrees of belief. Degrees of belief have an epistemic goal more immediate than graded accuracy, namely fitting the evidence. By fitting the total evidence, degrees of belief increase a rational ideal agent's expectation of their accuracy but the more immediate epistemic goal is fit with evidence. An agent controls fit with evidence but does not control graded accuracy because accuracy depends on the world. In a demon world with misleading evidence, rational degrees of belief are inaccurate and accurate degrees of belief are irrational. The overall

---

[24] Konek and Levinstein (2019) argue for pursuing accuracy in the right way. The qualification suggests an epistemic goal independent of accuracy. Schoenfield (2017) argues that, if accuracy were the primary epistemic goal, it would count against having imprecise credences. Granting this, if such credences are rational, it is because of an independent epistemic goal.

epistemic goal of having rational degrees of belief requires fit with evidence rather than accuracy.

Consider a doxastic domain having p as its only atom. Suppose that although the evidence an agent possesses yields a degree of belief assignment according to which $B(p) = 0.4$ and $B(\sim p) = 0.6$, according to the agent's assignment $B(p) = 0.4$ and $B(\sim p) = 0.4$. Suppose that Brier scores measure inaccuracy.[25] The agent's assignment at the p-world has a score of $(1 - 0.4)^2 + (0 - 0.4)^2 = 0.52$ and has the same score at the $\sim p$-world. The agent's switching to the probabilistic assignment $B(p) = 0.5$ and $B(\sim p) = 0.5$ is guaranteed to increase accuracy because at the p-world the assignment has the score $(1 - 0.5)^2 + (0 - 0.5)^2 = 0.5$ and has the same score at the $\sim p$-world. However, this probabilistic assignment does not fit the evidence. Moreover, the assignment supported by the evidence, with $B(p) = 0.4$ and $B(\sim p) = 0.6$, is not guaranteed to be more accurate than the agent's assignment, with $B(p) = 0.4$ and $B(\sim p) = 0.4$. The evidentially supported assignment at the p-world has the score $(1 - 0.4)^2 + (0 - 0.6)^2 = 0.72$ and at the $\sim p$-world has the score $(0 - 0.4)^2 + (1 - 0.6)^2 = 0.32$. Switching to an assignment guaranteed to increase accuracy may not be as epistemically desirable as switching to one that does not guarantee an increase in accuracy but fits the evidence.[26]

An argument for probabilism may look for something about degrees of belief themselves that makes it rational for them to comply with the axioms. It may seek noninstrumental reasons for compliance (expecting to find them because the intuition that the axioms are normative for degrees of belief does not rest on anything outside the realm of degrees of belief). It may try to identify an attainable standard of epistemic rationality, more immediate than promotion of graded accuracy, that imposes the requirement of compliance with the probability axioms.

A standard that fits the bill is having degrees of belief that fit the evidence. Weirich (2012) presents an argument for probabilism that uses rationality's requirement that degrees of belief match strengths of evidence. In an ideal agent, degrees of belief, if rational, inherit their additivity from the additivity of strengths of evidence.[27] Compliance with the probability axioms emerges from

---

[25] According to a common definition, the Brier score of an assignment at a world calculates for each proposition of the assignment the difference between the proposition's truth value (1 or 0) at the world and the proposition's degree of belief, squares the difference, and then sums the squared differences for all the propositions. In the example, an assignment covers the propositions p, $\sim p$, $(p \lor \sim p)$, and $(p \mathbin{\&} \sim p)$. Each assignment accurately assigns 1 to $(p \lor \sim p)$ and 0 to $(p \mathbin{\&} \sim p)$. Hence, the measure of an assignment's inaccuracy considers only p and $\sim p$.

[26] Easwaran and Fitelson (2012) make a similar point.

[27] Dempster–Shafer measures of evidence, described in Shafer (1976), are not additive. The argument sketched assumes an additive measure of strength of evidence such as Carnap

meeting the epistemic goal of fit with the evidence. An agent's internal reasons for degrees of belief lead to degrees of belief that comply with the axioms.

This argument, although epistemic and noninstrumental, rests on the assumption that strengths of evidence comply with the probability axioms and so is derivative.[28] Probabilism may seek support that is nonderivative and uses the nature of degrees of belief and norms for them without relying on the additivity of other quantities such as strengths of evidence. The norm of matching strengths of evidence yields compliance with the axioms but does not yield the axioms using only doxastic norms; it uses the non-doxastic fact that strengths of evidence are additive.

This section constructs an epistemic argument for probabilism that is non-instrumental and nonderivative. The argument shows that complying with the probability axioms is intrinsically good from an epistemic perspective, rather than arguing that compliance emerges from meeting epistemic goals involving factors besides degrees of belief. It shows that, from the epistemic perspective of rationality for doxastic attitudes, compliance with the probability axioms is intrinsically good, and not just instrumentally good, and not just derivative from epistemically desirable relations of degrees of belief with other quantities.

Following Ramsey (1931: 84), Skyrms (1987: 227), and Armendt (1992: 218), this section's argument for probabilism uses coherence. The argument shows that degrees of belief complying with the probability axioms have a form of coherence that is intrinsically good from the perspective of doxastic rationality. Coherence is an intrinsic property of a system of degrees of belief, whereas fit with evidence and graded accuracy are properties that degrees of belief have in relation to factors besides degrees of belief.

As Section 2 maintains, having a degree of belief is primarily a passive state and not an act. It is a response to evidence. A degree of belief is irrational if it responds badly to evidence, even if it has good consequences. Because of the passivity of a degree of belief, rationality's evaluation of it, unlike its evaluation of an act, does not consider the instrumental value of having the degree of belief. This section's argument for probabilism takes coherence as intrinsically good and, although coherence emerges from fit with evidence, relies on coherence rather than fit with evidence.

Consistency of beliefs is an intrinsically good property of beliefs, even if it derives from beliefs fitting evidence. Similarly, coherence is an intrinsically

---

(1962) advocates. Titelbaum (forthcoming: sec. 14.3) reviews the Dempster–Shafer theory of evidence and presents some problems it faces.

[28] Similarly, de-pragmatized sure-loss arguments for probability are derivative because they use the additivity of fair betting quotients to support the additivity of degrees of belief, taken as commitments to beliefs about fair betting quotients.

good property of degrees of belief, even if it derives from degrees of belief fitting evidence. Coherence is not only necessary for obtaining fit with evidence but is intrinsically good for degrees of belief from the perspective of doxastic rationality; it is intrinsically good from this perspective for degrees of belief to be coherent. Although coherence follows from fit with evidence, it has its own epistemic value. Suppose that an agent's degrees of belief fail to fit her evidence. Rationality nonetheless requires that they comply with the probability axioms because of the intrinsic good of the coherence that compliance entails.

An axiomatization of probability formulates truths from which all and only the laws of probability follow in an efficient way. The axioms of probability offer an efficient derivation of the laws of probability rather than an explanation of the laws of probability. In particular, the axioms do not provide any explanation of their own status as requirements of rationality. An explanatory argument for probabilism explains the normative status of the axioms and thereby the normative status of the laws of probability. The derivational project of the axioms differs from the project of explaining the norm of compliance with the laws of probability. For example, that logically equivalent propositions have equal degrees of belief is a consequence of compliance with the axioms, and inherits an explanation of them, but also has an independent explanation using coherence; assigning different degrees of belief to logically equivalent propositions is incoherent. A law deriving from the axioms may explain an axiom from which it derives. The requirement of equal degrees of belief for equivalent propositions may support the norm of compliance with the axioms, even if the requirement has a derivation from the axioms.

This section's argument for probabilism, besides being noninstrumental and nonderivative, explains why rationality requires compliance with the probability axioms. An explanatory argument for compliance with the axioms appeals to norms for degrees of belief that are explanatorily more basic than the axioms (although they do not replace the axioms as efficient means of deriving the laws of probability). An explanatory argument derives the norm of compliance with the probability axioms from more basic norms. To be explanatory, the argument must be general; it cannot make restrictive assumptions. It must identify features of degrees of belief that explain the rationality of compliance with the axioms in all cases meeting probabilism's assumptions and not just relations of degrees of belief to other things that motivate compliance with the axioms only when the relations hold. An explanatory argument, for generality, may use intrinsic features of a system of degrees of belief, such as coherence, to show the intrinsic value of compliance with the probability axioms.

4.3 Preliminaries

An argument for probabilism may use different methods to support different axioms and, because the axioms are general, may use different methods to support different instances of a single axiom. Also, it may use some axioms already established to support others yet to be established. For example, it may establish normality and then use this axiom in the case for additivity.

A preliminary step of this section's argument for probabilism is showing that an assignment of degrees of belief should be a function. That is, the assignment must have for any proposition in an agent's doxastic domain a single degree of belief; the agent may not have two quantitative doxastic attitudes to the same proposition. The idealizations about agents and their situations build a case for this requirement of singularity. According to the idealizations, an agent fully understands the propositions that sentences (in her language) express. She knows when two sentences (the vehicles of her thoughts) express the same proposition and so, if rational, does not assign two degrees of belief to the same proposition. The idealizations put aside cases in which an agent by mistake assigns two degrees of belief to the same proposition because she incompletely understands it in two ways that suggest different degrees of belief.

Rationality's requirement that degrees of belief obey the axioms of probability depends partly on conventions for representing doxastic attitudes and partly on norms for the attitudes. The axioms of non-negativity and normality by convention set the upper and lower limits of degrees of belief; the conventional scale for degrees of belief makes them range from 0 to 1. Non-negativity is completely conventional, and normality's normative component is just that rationality requires maximum degree of belief for a tautology. Finishing the case for probabilism requires only an argument for this normative component and for additivity.

Rationality requires a cognitively ideal person to be maximally confident of a tautology, and more generally of any a priori proposition, because it is obvious to such an agent that the proposition is true. A rational and cognitively ideal agent has a doxastic attitude of maximum strength toward an a priori truth (such as a tautology) and has a doxastic attitude of minimum strength toward an a priori falsehood (such as a contradictory proposition). This simple argument completes the case for probabilism except for the requirement of additivity, which Subsection 4.5 justifies.

4.4 Necessary Conditions

Belief may be comparative as well as quantitative, and norms of coherence govern comparative belief. However, comparative norms do not suffice for establishing additivity. Numerical representations of rational doxastic comparisons that adopt the conventions of using non-negative numbers and using 1 for certainty are not limited to probabilistic representations, which comply with additivity. Comparative norms can do no more than doxastically order propositions. For any probabilistic representation of the order, a transformation by squaring yields a non-probabilistic representation that conforms with non-negativity and normality. Nothing mandates adopting a probabilistic representation of the doxastic order of propositions. For example, the requirement of additivity governs a doxastic domain with a single atomic proposition p, and no comparative constraints prevent the assignments $B(p) = 0.4$, $B(\sim p) = 0.4$, and $B(p \vee \sim p) = 1$.[29]

Let $p \sim_b q$ stand for an agent's having equivalent doxastic attitudes to proposition p and to proposition q. According to a comparative norm: $p \sim_b q$ if and only if $(p \vee r) \sim_b (q \vee r)$ when r is incompatible with both p and with q. According to the norm, in all cases, r makes the same doxastic difference between p and $(p \vee r)$ as between q and $(q \vee r)$. However, the comparative norm does not require additivity for a degree of belief function B. It does not require $B(r)$ to have a value such that $B(p \vee r) = B(p) + B(r)$ and such that $B(q \vee r) = B(q) + B(r)$. The comparative norm does not yield additivity. Support for additivity requires more than norms of comparative belief.

Support for a normative principle may show that nontrivial implications of the principle hold. This form of support is not conclusive but still carries some weight if the case for the implications is independent of the case for the norm. Support of this type for the additivity of rational degrees of belief shows that they meet necessary conditions of additivity. Although norms of comparative belief do not establish additivity, they establish some conditions that are necessary for additivity.[30]

---

[29] Suppose that a norm requires an assignment of degrees of belief over a finite doxastic domain to be embeddable in an assignment of degrees of belief over an infinite doxastic domain. Still, comparative norms do not yield probabilism because they do not require the assignment over the infinite domain to be probabilistic.

[30] A representation theorem for degrees of belief shows that if comparative belief meets norms such as transitivity and has a rich enough structure, then a unique probability function represents comparative belief. A probability function by definition satisfies the axioms of probability, so the norms of comparative belief do not establish that a rational ideal agent satisfies the axioms. However, norms of comparative belief ground a necessary condition of probabilistic degrees of belief, namely having comparative beliefs that a probability function may represent, assuming that degrees of belief agree with comparative belief. The norms offer support for having degrees of belief that comply with the probability axioms by supporting a necessary condition of compliance with the axioms. Although a representation theorem's conditions on comparative belief include non-normative structural conditions, such as the completeness of comparisons, these structural conditions are necessary only for the uniqueness of the representation of comparative belief by a probability function. They are not necessary for establishing the existence of such a representation. The norms of comparative belief do this by themselves.

A norm of coherence requires that degrees of belief agree with comparative belief, taken in a technical sense that includes doxastic comparison of propositions disbelieved. Using $p \succeq_b q$ to represent comparative belief for a pair of propositions p and q, a norm of coherence requires that $B(p) \geq B(q)$ if and only if $p \succeq_b q$. Therefore, norms of comparative belief impose norms on comparisons of degrees of belief.

According to *compositionality* of the components of an exclusive disjunction – that is, a disjunction with logically exclusive disjuncts – the attitude to the disjunction is a function of the attitudes to the disjuncts. Compositionality assumes only a classification of doxastic attitudes as equivalent or not equivalent. It justifies a principle of interchange of one component of an exclusive disjunction with another component to which an agent has an equivalent doxastic attitude and such that an interchange also produces an exclusive disjunction. For example, if, concerning the day's weather, an agent has the same doxastic attitude to no rain and snow, ($\sim r$ & s), as to no rain and no snow, ($\sim r$ & $\sim s$), then the agent has the same doxastic attitude to the disjunction ($r \lor (\sim r$ & $s)$) as to the disjunction ($r \lor (\sim r$ & $\sim s)$). The interchange of disjuncts works because the agent has the same doxastic attitude to the propositions interchanged.

Mutual *separability* of the components of exclusive disjunctions holds if and only if the ranking of the disjunctions according to doxastic attitudes agrees with the ranking of their first disjuncts when the second disjunct is fixed and, similarly, agrees with the ranking of second disjuncts when the first disjunct is fixed. That is, if p is incompatible with q and with r, then $(q \lor p) \succeq_b (r \lor p)$ if and only if $(q \succeq_b r)$, and $(p \lor q) \succeq_b (p \lor r)$ if and only if $(q \succeq_b r)$. Separability defined this way assumes only comparison of doxastic attitudes.

The compositionality and separability of the degrees of belief of the components of an exclusive disjunction are necessary for their additivity, and these features of degrees of belief follow from their counterparts for comparative belief, assuming agreement of comparative belief and degrees of belief. To support additivity, this section argues for the compositionality and separability of coherent comparative belief.

The reasons for doxastic attitudes to a pair of exclusive propositions are independent. Hence, for a rational ideal agent, the doxastic attitude to each proposition contributes independently to the doxastic attitude to their disjunction. Hence, the doxastic attitude to the disjunction is a function of the doxastic attitudes to the disjuncts; compositionality holds. Also, the doxastic ranking of disjunctions with a common disjunct agrees with the doxastic ranking of the propositions forming the disjunct that varies. Thus, separability obtains.

Let us consider separability in more detail. Suppose that a vector (a list) of two variables represents the structure of the composites in a set of composites and a vector of values of the two variables represents a composite. The first variable is *separable* from the second if and only if for all ways of fixing the value of the second variable the order of values of the first variable agrees with the order of vectors of values of both variables. The separability of the second variable from the first has a similar definition.[31]

Consider variables x and y such that the possible values of x are x_1 and x_2 and the possible values of y are y_1 and y_2. Suppose the order of vectors of the variables' possible values is: $(x_1, y_1) \prec (x_2, y_1) \prec (x_1, y_2) \prec (x_2, y_2)$. Also, suppose that values of the variables have the order of their subscripts, that is, $x_1 \prec x_2$ and $y_1 \prec y_2$. Then x is separable from y because the order of x's values agrees with the order of vectors of values of both variables for any fixed value of y. Similarly, y is separable from x.

The axiom of additivity treats an exclusive disjunction. Given that the degree of belief in an exclusive disjunction equals the sum of the degrees belief in the disjuncts, the degree of belief in the first disjunct is separable from the degree of belief in the second disjunct. No matter how the degree of belief in the second disjunct is fixed, increasing the degree of belief in the first disjunct increases the degree of belief in the disjunction. Similarly, the degree of belief in the second disjunct is separable from the degree of belief in the first disjunct. Ordering by degree of belief, the order of exclusive disjunctions with a common disjunct agrees with the order of the other disjuncts. For exclusive disjunctions, two variables having, respectively, degrees of belief in the two disjuncts as values, are separable from each other. Given agreement of comparative and quantitative belief, the separability of comparative belief for exclusive disjunctions is a necessary condition of the separability, and additivity, of degrees of belief. To support additivity, I elaborate support for the separability of comparative belief.

Consider a disjunction $(p \lor q)$ of exclusive propositions. The reasons for a doxastic attitude to p are independent of the reasons for a doxastic attitude to q. Hence, a variable for the first disjunct should be separable from a variable for the second disjunct. However, the variable for the second disjunct is fixed, the order of values of the variable for the first disjunct should agree with the order of disjunctions formed with the values of the variable for the first disjunct and the fixed value of the variable for the second disjunct. By symmetry, the variable for the second disjunct should also be separable from the variable for the first disjunct. Thus, the variables should be separable from each other.

---

[31] In general, this type of separability is a feature of subsets of a set of variables generating the components of composites, given a way of ordering the composites and subsets of their components. In a set of variables, a subset is *separable* from the others if and only if for all ways of fixing the values of the other variables, the order of subvectors of values of variables in the subset agrees with the order of vectors of values of all variables.

To confirm the separability of comparative belief, I show how for special sets of exclusive disjunctions separability follows from other norms of comparative belief concerning pairs of equivalent propositions and pairs of propositions such that one proposition entails the other.

Disjoining to an exclusive disjunction's first disjunct a proposition incompatible with both its disjuncts generates another exclusive disjunction. Consider a set of exclusive disjunctions extended this way. Suppose that comparative belief orders the disjunctions and their extended first disjuncts. The separability of the variable for first disjuncts from the variable for second disjuncts requires the order of the first disjuncts to agree with the order of the disjunctions given any way of fixing the second disjunct. For example, take the disjunctions $(p \lor q)$, $(p \lor r)$, $(s \lor q)$, $(s \lor r)$, letting s be the disjunction of p with a proposition incompatible with each of p, q, and r. If the variable for first disjuncts is separable from the variable for second disjuncts, and less belief is invested in the proposition p than in the proposition s, then less belief is invested in $(p \lor q)$ than in $(s \lor q)$ and less belief is invested in $(p \lor r)$ than in $(s \lor r)$. That is, if $p \prec_b s$, then $(p \lor q) \prec_b (s \lor q)$ and $(p \lor r) \prec_b (s \lor r)$.

In the set of disjunctions involving s, let s be $(p \lor \sim(p \lor q \lor r))$. The extended set of exclusive disjunctions then has these members:

$(p \lor q)$

$(p \lor r)$

$((p \lor \sim(p \lor q \lor r)) \lor q)$

$((p \lor \sim(p \lor q \lor r)) \lor r)$

A norm of coherence for comparative belief requires that an ideal agent invest no more belief in a proposition than in any proposition it entails. Consider the special case of p and $(p \lor q)$, with p and q being incompatible. The norm requires that $p \preccurlyeq_b (p \lor q)$. Coherence requires the inequality to be strict if the agent invests more belief in q than in a contradictory proposition, such as $(p \ \& \ \sim p)$. In this case, coherence requires that $p \prec_b (p \lor q)$. This norm of comparative belief is, I assume, explanatorily more basic than the norm of separability for comparative belief and so may contribute to its explanation.

Suppose that in the extended set of exclusive disjunctions, the agent invests more belief in $\sim(p \lor q \lor r)$ than in a contradictory proposition. Then the norm concerning entailment of a disjunction requires that

$p \prec_b (p \lor \sim(p \lor q \lor r))$

$(p \lor q) \prec_b ((p \lor q) \lor \sim(p \lor q \lor r))$

$(p \lor r) \prec_b ((p \lor r) \lor \sim(p \lor q \lor r))$

Because coherence requires equal belief in equivalent propositions, because $((p \lor \sim(p \lor q \lor r)) \lor q)$ is equivalent to $((p \lor q) \lor \sim(p \lor q \lor r))$, and because $((p \lor \sim(p \lor q \lor r)) \lor r)$ is equivalent to $((p \lor r) \lor \sim(p \lor q \lor r))$, coherence also requires that

$p \prec_b (p \lor \sim(p \lor q \lor r))$

$(p \lor q) \prec_b ((p \lor \sim(p \lor q \lor r)) \lor q)$

$(p \lor r) \prec_b ((p \lor \sim(p \lor q \lor r)) \lor r)$

Hence, assuming coherence, the order of p and $(p \lor \sim(p \lor q \lor r))$ agrees with the order of the disjunctions containing them as first disjunct, fixing the second disjunct either as q or as r. Therefore, the variable for first disjuncts is separable from the variable for second disjuncts. For certain sets of exclusive disjunctions, the norm for entailments supports the norm of separability. This support for separability, although limited to special cases, illustrates a way of using norms of coherence for comparative belief to explain why separability is a norm of coherence for comparative belief.

Any separable function may represent the separability of comparative belief but not all separable functions are adequate representations given that degrees of belief satisfy the axioms of non-negativity and normality and, for a rational ideal agent, agree with comparative belief. Using multiplication to represent separability generates incoherence. For example, assume that degrees of belief are non-negative and that degrees of belief in tautologies equal 1 and let p and $\sim p$ enjoy equal degrees of belief. Then, so that $B(p \lor \sim p)$, as represented by $B(p) \times B(\sim p)$, equals 1, $B(p)$ and $B(\sim p)$ must each equal 1. However, an ideal agent should not assign both a proposition and its negation the same degree of belief that she assigns a tautology. Moreover, multiplication does not provide a separable function if one variable may have the value 0. In this case, the order of composites does not agree with the order of the other variable's values. The composites all have the same ranking although the other variable's values do not. A multiplicative representation of the separability of comparative belief within a set of exclusive disjunctions must exclude cases in which an agent invests equal belief in a contradictory proposition and in some disjunct. Hence, addition represents comparative belief's separability better than does multiplication.[32]

---

[32] I presented the ideas in this section at a 2014 University of Utah conference entitled "Inductive Logic and Confirmation in Science II" and am grateful for comments received there.

4.5 Additivity

General additivity for a quantity, such as length, governs a particular two-part composite and also any composite formed by altering one of the components. The length of a straight stick is its contribution to the length of a two-part composite straight stick to which it belongs. Moreover, the length of a straight stick makes a constant contribution to two-part composite straight sticks to which it belongs as the other component varies. The constant contribution of the stick's length to the lengths of two-part composites to which it belongs makes length additive. The additivity of weight has a similar justification. The constant contribution of an object's weight to two-part composites to which it belongs, as the other component varies, makes weight additive. This section advances this type of justification of additivity for degrees of belief, assuming that they are coherent, as they are for a rational ideal agent.

Consider a disjunction of two exclusive propositions. The justification of additivity for rational degrees of belief shows that the degree of belief in a disjunct makes a contribution to the degree of belief in the disjunction. Then it shows that the contribution is constant as the other disjunct varies. These two points establish additivity. Because the degree of belief in a proposition makes a constant contribution to the degrees of belief in exclusive disjunctions to which the proposition belongs, degree of belief is additive. This type of argument for additivity may use coherence constraints exclusively and so show, from the epistemic perspective of doxastic rationality, that additivity is an intrinsically good property of an assignment of degrees of belief.

Rationality imposes constraints on differences between the doxastic attitudes that degrees of belief represent and so on degrees of belief. The difference between a degree of belief that a disjunction of exclusive propositions holds and a degree of belief that a disjunct holds should be non-negative. A positive degree of belief that a disjunct holds should make a contribution to the degree of belief that the disjunction holds. Also, its contribution should be independent of the other disjunct. The independence of reasons for doxastic attitudes to exclusive propositions supports these norms.

I use marginal degree of belief, MB, taken as a primitive, to state two coherence requirements for degrees of belief. A marginal degree of belief represents a quantitative doxastic comparison of a proposition and its disjunction with another proposition. By convention, marginal degrees of belief are on the same scale as degrees of belief. Suppose that a disjunction $(p \lor q)$ is a composite built by starting with p and then disjoining q. The *marginal degree of belief* of q in its disjunction with p is the contribution q makes to the degree of belief that $(p \lor q)$. $MB(q|p)$ is the increment in degree of belief as the agent

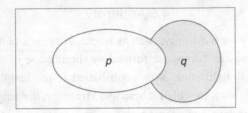

Figure 1 Marginal degree of belief of q given p

moves from p to $(p \lor q)$. In the Venn diagram of Figure 1, the representations of p and of q are ovals, with part of the oval representing q occluded by the oval representing p. The shaded area of the diagram represents $MB(q|p)$. It equals the area representing $(q \ \& \sim p)$.

Marginal degree of belief has a manifestation in behavior concerning gambles. In typical circumstances, for a rational ideal agent with linear utility of money, $MB(q|p)$ equals the largest percent of a dollar that the agent is willing to pay to transform a gamble that pays a dollar if p and nothing otherwise into a gamble that pays a dollar if $(p \lor q)$ and nothing otherwise.

An explanatory argument for the requirement of additivity for degrees of belief digs deeper than the requirement itself. Because the requirement does not mention marginal degrees of belief, a way to dig deeper, without leaving the realm of degrees of belief, is to use marginal degrees of belief to explain the requirement. This section's explanatory argument obtains the norm of additivity from two simpler norms of coherence involving marginal degrees of belief.

According to a norm of coherence, $MB(q|p)$ equals the difference between $B(p \lor q)$ and $B(p)$, given that degrees of belief exist for p and for $(p \lor q)$. That is, $MB(q|p) = B(p \lor q) - B(p)$. For example, the marginal degree of belief that a die roll yields 2, in a disjunction built starting with the roll's yielding a 1, equals the difference between the degree of belief that the roll yields 1 or 2 and the degree of belief that it yields 1. Assuming the usual degrees of belief about outcomes of the roll, the marginal degree of belief is $\frac{2}{6} - \frac{1}{6} = \frac{1}{6}$. The general equation holds not by the definition of marginal degree of belief but because of a requirement of coherence. I call the equation *the principle of contribution*.

Marginal degree of belief assumes that a disjunction has an order of composition. It assumes that one disjunct is first in construction of the disjunction. The order of construction may reflect an agent's attention moving from one disjunct to the disjunction.

A proposition's marginal degree of belief may depend on whether it comes first or second in construction of a nonexclusive disjunction. For example, $B(p \lor q) = B(p) + MB(q|p)$, assuming that p comes first in the composition of $(p \lor q)$, so that $B(p)$ equals the contribution of the degree of belief that p to the degree of belief that $(p \vee q)$. If q is first in the composition of $(p \vee q)$, then $B(q)$ equals the contribution of the degree of belief that q to the degree of belief that $(p \vee q)$. Hence $B(p \vee q) = B(q) + MB(p|q)$. Comparing $B(q)$ and $MB(q|p)$ compares the contribution of the degree of belief that q when it comes first and when it comes second in the composition of $(p \vee q)$. Possibly, $B(q) \neq MB(q|p)$. Then q's marginal contribution depends on the order of composition of $(p \vee q)$. However, if q is incompatible with p, then, for a rational ideal agent, q's marginal contribution is independent of the order of composition of $(p \vee q)$, so $B(q) = MB(q|p)$. Also, p's marginal contribution given q, $MB(p|q)$, is independent of the order of composition of $(p \vee q)$, so $B(p) = MB(p|q)$. Because p and q are incompatible, the order of composition of $(p \vee q)$ does not affect either's marginal degree of belief given the other.

To elaborate, consider a disjunction of exclusive propositions p and q. $MB(q|p)$ equals the contribution of the degree of belief that q to the degree of belief that $(p \vee q)$ given q's disjunction with p to form $(p \vee q)$. An agent begins with a degree of belief that p and ends with a degree of belief that $(p \vee q)$, and $MB(q|p)$ equals the contribution of the degree of belief that q to the degree of belief that $(p \vee q)$. The shaded area of the Venn diagram in Figure 2 illustrates $MB(q|p)$.

Assuming degrees of belief for all the propositions of a doxastic domain, for any proposition p in the domain and all q in the domain incompatible with p, $B(q) = MB(q|p)$. That is, the degree of belief that q equals the marginal degree of belief of q given p. This equality states a requirement of coherence. I call the equality *the principle of independence*.

Because the principle of independence, $B(q) = MB(q|p)$, is a generalization, it assumes that $B(q)$ does not vary as p varies among propositions incompatible with q. This assumption holds because incompatible propositions have independent truth conditions. Consequently, the degree of belief that p and the degree of belief that q have independent effects on the degree of belief that $(p \vee q)$. If p and q are incompatible, the reasons for the degree of belief that q are

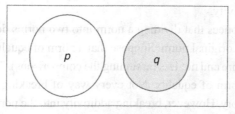

Figure 2 Marginal degree of belief of q given p for exclusive propositions p and q

independent of the reasons for the degree of belief that p. An ideal agent, if rational, assigns the same degree of belief to a proposition given its disjunction with any incompatible proposition; the disjunction of the two propositions is *nonproductive* in the sense that the disjunction entails only propositions that some disjunct entails. Hence, for a rational ideal agent, the contribution that the degree of belief that q makes to the degree of belief that $(p \lor q)$ is the same for all p incompatible with q. Hence, the marginal degree of belief of q given p, $MB(q|p)$ – that is, the contribution of the degree of belief that q to the degree of belief that $(p \lor q)$ – is constant as p varies among propositions incompatible with q. Rationality's constraints on an ideal agent's doxastic attitudes make a disjunct's degree of belief contributionally independent of the other disjunct. A disjunct's degree of belief makes a constant contribution to the degree of belief in any disjunction with an incompatible proposition. Given the independence of the reasons for the doxastic attitudes to p and to any incompatible proposition, an incompatible proposition disjoined to p does not affect the degree of belief that p. The degree of belief that p is independent of the degree of belief that the other proposition holds.

This section adopts the norm of contribution and the norm of independence for marginal degrees of belief and draws on the two norms to support the norm of additivity for an ideal agent's degrees of belief. The two norms together yield additivity for a disjunction of two logically exclusive propositions, assuming that the propositions have degrees of belief.

According to the norm of contribution, $MB(q|p) = B(p \lor q) - B(p)$. If p and q are exclusive, then, according to the norm of independence, $B(q) = MB(q|p)$. Hence, by transitivity of identity, $B(q) = B(p \lor q) - B(p)$. Therefore $B(p \lor q) = B(p) + B(q)$. Degrees of belief are additive given their compliance with the norm of contribution and the norm of independence.

Although the norm of contribution is closely related to addition, the norm of additivity derives from two norms, contribution and independence, taken together. Neither by itself suffices. The norm of contribution and the norm of independence directly yield the norm of additivity but nonetheless explain the norm of additivity because, as I assume, they are explanatorily more basic than the norm of additivity.

An objection notices that dividing a norm into two norms does not ensure an explanation of the original norm. Suppose that a norm of equality is divided into two norms: not more and not less (assuming the comparisons). This division need not explain the norm of equality. Not every way of breaking a norm into two norms is explanatory. However, breaking additivity into the norm of contribution and the norm of independence yields an explanation of additivity, even if deeper-going explanations are also possible.

From the outset, I assumed that probabilism is true and have just explored ways of supporting its truth. This section's argument for additivity does not have the job of using something more persuasive than additivity to convince doubters of additivity. Its job is to give epistemic, noninstrumental, nonderivative, and explanatory reasons why additivity holds. The argument shows that additivity follows from two norms of coherence involving marginal degrees of belief. Its explanation of additivity completes the type of support sought for probabilism.[33]

5 The Expected-Utility Principle

Besides norms for degrees of belief, rationality imposes norms for degrees of desire. This section treats decision problems in which a rational ideal agent, having sufficient experience and information, assigns degrees of belief to the possible outcomes of options and degrees of desire to options and the possible outcomes of options. The degrees of belief constitute probabilities, that is, rational degrees of belief, and the degrees of desire constitute utilities, that is, rational degrees of desire (all things considered). Section 4 explains why probabilities so defined conform to the laws of probability, and this section explains why utilities so defined conform to the expected-utility principle. To simplify terminology, instead of arguing that rationality requires degrees of desire to comply with the principle, it assumes that degrees of desire are rational, and hence are utilities, and argues that utilities conform to the principle.[34]

5.1 Expected Utility

Taking the options in a decision problem to be acts, the expected-utility principle states that an act's utility, calculated using a finite partition of its possible outcomes given its realization, is a probability-weighted sum of the utilities of the possible outcomes, assuming that the utilities of the possible outcomes are finite. According to the principle, the utility of a bet is the probability of winning times the utility of winning, plus the probability of

---

[33] I presented the ideas in this section at a 2014 *Maisons des Sciences Economique* conference entitled "Economics and Psychology in Historical Perspective" and am grateful for comments received there.

[34] Utility, as rational degree of desire in an ideal agent, is responsive to values that rationality recognizes. Rationality's standards of evaluation apply to doxastic attitudes, conative attitudes, and acts; and meeting its standards has value. However, I do not assume that rationality requires an agent to have as a goal meeting these standards. I leave open whether rationality requires an agent to want to be rational. I do not assume that meeting rationality's standards of evaluation for acts motivates an agent's acts, only that the rationality of the agent's acts depends on meeting the standards.

losing times the utility of losing. This version of the principle does not use possible states, as intermediaries, to obtain an act's possible outcomes.

I adopt causal decision theory's version of the expected-utility principle and so, as the probability of a possible outcome of an act, use the probability of the conditional that if the act were performed the outcome would obtain. However, this section's support for the expected-utility principle puts aside the distinction between causal decision theory and evidential decision theory, which uses the probability of the outcome given the act, by addressing only decision problems in which, for every possible outcome of an act, its probability conditional on the act equals the probability of the corresponding conditional. General support for the expected-utility principle dispenses with this restriction.

The expected-utility principle assumes probabilism and hence that in a rational ideal agent the degrees of belief grounding informationally derived degrees of desire comply with the axioms of probability. For example, it assumes that an agent assigns a tautology the maximum degree of belief, 1. Otherwise, the expected-utility principle might conclude that the utility of an act that with certainty yields some possible outcome differs from the utility of that possible outcome. Section 4 lays the groundwork for the expected-utility principle by supporting probabilism.

The expected-utility principle also assumes that an act's possible outcomes are comprehensive; they include all features the agent cares about. They are not just monetary outcomes if the agent cares about more than money. The principle would require an agent choosing between job offers to select the job paying the most if it neglected nonmonetary considerations, such as job satisfaction. However, selecting the best-paying job is irrational if another job offered would bring satisfaction that in the agent's eyes more than compensates for its not maximizing pay.

Some theorists, for example Buchak (2013), advance Allais's (1953) paradox as an objection to the expected-utility principle. In a simple version of the paradox, an agent forms a preference between \$3,000 and a gamble with a $\frac{4}{5}$ chance of paying \$4,000, and also forms a preference between a gamble with a $\frac{1}{4}$ chance of paying \$3,000 and a $\frac{1}{5}$ chance of paying \$4,000. A typical agent prefers the sure \$3,000 and prefers the $\frac{1}{5}$ chance of \$4,000. If possible outcomes are just amounts of money, and U stands for utility, with no monetary gain or loss serving as the zero point, the first preference reveals that $U(\$,3000) > \frac{4}{5} U(\$4,000)$ and the second preference reveals that $\frac{1}{4} U(\$3,000) < \frac{1}{5} U(\$4,000)$. Multiplying the sides of the second inequality by 4 shows that the inequalities are inconsistent.[35]

---

[35] This version of the paradox relies on the common ratio effect rather than the common consequence effect.

Taken as an objection to the expected-utility principle, the paradox assumes that the typical preferences are rational. To justify their rationality, it assumes that the agent has a rational aversion to an act's risk in the sense of the variability of the utilities of the act's possible outcomes. Given this assumption, a comprehensive possible outcome of an act with possible outcomes of variable utility includes the act's risk, not just money, so that utilities of the act's possible outcomes register the agent's aversion to the act's risk and the inequalities the preferences express become consistent. Taking outcomes comprehensively to include an option's risk eliminates the objection to the expected-utility principle.[36]

To handle rational aversion to risk, some decision principles evaluate an act using an agent's attitude to risk as a third factor joining the probabilities and utilities of possible outcomes. However, to give this third factor some structure, these decision principles overly restrict an agent's attitude to risk. For example, Buchak's (2013) principle requiring maximization of risk-weighted expected utility prohibits an agent's being indifferent to small risks and averse to large risks.[37] As Weirich (2020: sec 5.6) argues, it constrains aversion to risk more than rationality does.

Rationality is permissive about matters of taste, such as evaluation of eating peaches and cream. The utility of a meal absorbs evaluation of this dessert if it is part of the meal. The utility of an act's possible outcome similarly absorbs evaluation of the act's risk, granting that an act's risk is part of its possible outcome. Because rationality is permissive about attitudes to risks, a decision principle does best to count an act's risk as part of the act's possible outcomes and then let the utility of a possible outcome evaluate together all the possible outcome's parts. The agent's evaluation of a possible outcome handles the agent's evaluation of the act's risk without imposing unwarranted restrictions on the agent's attitude to the risk or on the way the agent uses her attitude to evaluate a possible outcome. No advantage comes from making risk an explicit

---

[36] Hammond (1988) endorses taking consequences comprehensively. A common objection to this view of consequences claims that it makes maximization of expected utility trivial. The objection imagines using features of outcomes as fudge factors to stave off counterexamples to the rationality of expected-utility maximization. However, only features of outcomes that matter to the agent affect expected utility. This restriction ensures that taking outcomes comprehensively does not trivialize expected-utility maximization, as Weirich (2015a: sec. 2.4) explains. Baccelli and Mongin (2020) present an objection to the restriction but the idealizations of the decision model I construct block their objection.

[37] The risk-weights depend only on the probabilities of possible outcomes, not their utilities. Hence Buchak's principle makes the weights insensitive to the size of stakes. Suppose that an agent cares only about risk and chances for money and assigns linear utilities to amounts of money. Then, according to Buchak's principle, if the agent is willing to bet $1 at even odds that a coin toss will yield heads, the agent must be willing to bet $1,000 at even odds that the coin toss will yield heads.

third factor in a decision principle because rationality does not strictly regulate attitudes to risks, or the attitudes' role in choice, to constrain decisions.

5.2 Justifying the Expected-Utility Principle

An argument for the expected-utility principle, which Weirich (2020: chap. 4) presents in detail, uses the distinction between a proposition's utility and its intrinsic utility. A proposition's utility evaluates the comprehensive outcome of the proposition's realization. In contrast, a proposition's *intrinsic utility* evaluates only the (a priori) implications of the proposition's realization. It omits, for example, causal consequences of the proposition's realization because they are not implications of the proposition's realization. A consequence of buying a diamond is possession of a valuable jewel. However, buying a diamond does not imply possession of a valuable jewel. In a world that does not value diamonds, buying one does not bring possession of a valuable jewel.

For an agent, the intrinsic utility of a possible world equals the utility of the possible world (assuming coordination of the scale for utility and the scale for intrinsic utility) because the possible world, taken as a consistent proposition specifying all that matters to the agent, implies all its features. This equality is a bridge between utility and intrinsic utility. Given complete information, an act's utility equals the intrinsic utility of the world that results from the act's realization, that is, the act's comprehensive outcome.

Other equalities also connect utility and intrinsic utility. Consider the proposition that an event with such and such a utility occurs. The proposition expressing the event's utility has an intrinsic utility equal to the event's utility. Take the proposition that an event of utility 3 occurs. The intrinsic utility of the proposition equals 3. Also, suppose that an event has an evidential probability as well as a utility. The intrinsic utility of the (evidential) chance for the event's realization equals the probability of the event times the utility of a proposition expressing the event's utility. For example, if an event has a probability of ½ and a utility of 3, then the intrinsic utility of the chance for the event, characterized as an event with a utility of 3, equals ½ × 3. In the special case that the chance of an event's realization equals 1, the intrinsic utility of the chance equals the utility of the event as characterized by its utility.

An argument for the expected-utility principle begins with the principle that the intrinsic utility of a chance for a possible outcome equals the probability-weighted utility of the possible outcome, assuming a characterization of the chance as an evidential probability of an event having the utility of the possible outcome. Let U stand for utility, IU stand for intrinsic utility, and $ch(P(o_i), U(o_i))$ stand for the chance of an outcome o_i characterized using the outcome's probability and its utility, that is, as a probability of an event of such and such utility. Then the principle states that $IU(ch(P(o_i),U(o_i))) = P(o_i)U(o_i)$. According to the principle, the greater the probability of the outcome, the greater the intrinsic utility of the chance; and the lower the probability of the outcome, the lower the intrinsic utility of the chance.

Next, the argument asserts that the intrinsic utilities of (evidential) chances for an act's possible outcomes, which form a partition given the act, are (finitely) additive. Given two (exclusive) possible outcomes, the intrinsic utility of having the chance for the disjunction of the possible outcomes equals the sum of the intrinsic utility of the chance of having the first outcome and the intrinsic utility of the chance of having the second outcome. This pairwise additivity holds because each chance is independent of the other and makes a constant contribution to the intrinsic utility of any pairing with another chance.

Suppose that an agent chooses to eat an apple, a pear, or a banana by random selection. Each possible outcome has a $\frac{1}{3}$ chance of realization. The intrinsic utility of the chance of eating a pear or eating a banana equals the sum of the intrinsic utility of the chance of eating a pear and the intrinsic utility of the chance of eating a banana. If the intrinsic utility of each possible outcome equals 1, then the intrinsic utility of the chance of their disjunction equals $\frac{2}{3}$.

Applying iteratively the principle of pairwise additivity yields that the intrinsic utility of the chance for the disjunction of an act's possible outcomes equals the sum of all the intrinsic utilities of the chances for the act's possible outcomes. That is, it equals $\sum IU(ch(P(o_i),U(o_i)))$; or replacing the intrinsic utilities of the chances with the equivalent probability-utility products, it equals $\sum P(o_i)U(o_i)$. In the example, this formula yields 1 as the intrinsic utility of selecting a fruit by randomization.

The probability-utility product for the disjunction of an act's possible outcomes equals 1 times the act's utility because the act produces the disjunction with certainty, and the utility of the disjunction equals the act's utility, as the disjunction of the act's possible outcomes is equivalent to the act under the assumption that the possible outcomes are comprehensive and include the act.[38] Hence, the intrinsic utility of the chance for the disjunction of the act's possible outcomes, besides equaling $\sum P(o_i)U(o_i)$, also equals the act's utility, that is, $U(a)$, letting a stand for the act. By the symmetry and transitivity of identity, $U(a) = \sum P(o_i)U(o_i)$, which is the expected-utility principle.

---

[38] A typical act has many consequences. Some it produces with certainty and others without certainty. It produces the disjunction of its exclusive and exhaustive possible outcomes with certainty and it also produces a particular disjunct but without certainty. Jeffrey ([1965] 1990) takes an act to produce a disjunction of possible outcomes, taking a possible outcome to be a conjunction of the act and a state from a partition of states of the world.

To illustrate, take an act a with just two (exclusive) possible outcomes o_1 and o_2. The chance the act offers is certainty of the disjunction of its possible outcomes, assuming that the possible outcomes are comprehensive. Applying the probability-utility formula for the intrinsic utility of a chance, the intrinsic utility of this chance equals $1 \times U(o_1 \vee o_2)$. Because $U(a) = U(o_1 \vee o_2)$, the intrinsic utility of the chance also equals $U(a)$. Applying the additivity of intrinsic utilities of chances, the intrinsic utility of the chance the act offers also equals the sum of the intrinsic utilities of the chances for o_1 and for o_2. That is, it equals $IU(ch(P(o_1),U(o_1))) + IU(ch(P(o_2),U(o_2)))$. Hence, by the symmetry and transitivity of identity, $U(a) = IU(ch(P(o_1),U(o_1))) + IU(ch(P(o_2),U(o_2)))$. Therefore, applying the probability-utility formula for the intrinsic utility of a chance to the summands, $U(a) = P(o_1)U(o_1) + P(o_2)U(o_2)$, as the expected-utility principle states.

This section and Section 4 argue for norms governing quantitative doxastic and conative attitudes. Section 6 extends these norms to nonquantitative doxastic and conative attitudes.

6 Norms for Imprecise Attitudes

An agent's doxastic attitude targets a proposition, and an agent's doxastic state comprises her doxastic attitudes to all the propositions in the doxastic domain she adopts. Similarly, an agent's conative attitude targets a proposition, and in a decision problem an agent's conative state comprises her conative attitudes to all the propositions in the *decision frame* she adopts. For a decision problem, an agent's decision frame comprises propositions expressing her options and the possible outcomes of the options. A resolution of an agent's decision problem issues from a doxastic and conative state that combines her doxastic and conative attitudes to propositions of her decision frame.

In cases of imprecision, a set of degree of belief assignments represents an agent's doxastic state, and a set of degree of desire assignments represents her conative state. A set of *representatives*, or pairs of a degree of belief assignment and a degree of desire assignment, represents the combination of her doxastic state and her conative state. The representation attributes to the combined state the shared features of representatives relevant to rational choice. Accordingly, the state has the shared features of representatives pertaining to comparisons of options, including doxastic comparisons of possible outcomes of options. In particular, if all the representatives rank one option higher than another, then so does the agent's doxastic and conative state.

A rational ideal agent may not have a degree of belief assignment and a degree of desire because of insufficient evidence and experience. However, if the agent has such assignments, familiar standards of rationality apply. Sections 4 and 5 argue that, for an ideal agent, a rational degree of belief assignment is a probability function and a rational degree of desire assignment is a utility function (and so observes the expected-utility principle). This section argues that, for an ideal agent, these standards of rationality extend to the degree of belief assignments and the degree of desire assignments appearing in the pairs of such assignments that together represent a nonquantitative doxastic and conative state. The degree of belief assignments must be probability functions, and the degree of desire assignments must be utility functions. Sections 4 and 5, by explaining norms for degrees of belief and degrees of desire, ground an explanation of norms for pairs of a degree of belief assignment and a degree of desire assignment in a set representing an agent's doxastic and conative state.

6.1 Imprecise Probabilities

Rationality allows gaps in an ideal agent's assignment of degrees of belief to the propositions of a doxastic domain. Such gaps create imprecision. Because of such gaps, and other sources of imprecision, a representation of an agent's doxastic state may use a set of degree of belief assignments rather than a single assignment.

If a set of degree of belief assignments that represents an ideal agent's doxastic state has an assignment that is not a probability function, then the assignment may yield irrational decisions, despite the agent's following the permissive principle of choice that Section 7 advances. In particular, using a degree of belief assignment that does not comply with the probability axioms may lead to a system of bets that creates a sure loss. For example, if an assignment does not accord maximal degree of belief 1 to a tautology, then it permits wagering money against the tautology despite the patent inevitability of losing the wager. The pragmatic argument for an agent's having a doxastic state represented by a set of probabilistic degree of belief assignments justifies the norm. Nonetheless, an alternative argument may offer an epistemic explanation of the norm.

One argumentative approach takes each degree of belief assignment in a set representing an ideal agent's doxastic state to represent a hypothetical state of mind that the agent might have if the agent were to have a precise assignment of degrees of belief.[39] The hypothetical state is rational only if the degree of belief assignment is a probability function. However, a requirement for such a hypothetical state of mind does not clearly yield a requirement for an agent's

---

[39] Joyce (2010) suggests thinking of each degree of belief assignment as the assignment of a member of a committee representing the agent. Taking each committee member's assignment as a hypothetical assignment of the agent yields this view of the representative assignments.

actual state of mind and the degree of belief assignments in the set that represents it.

Another approach maintains that a doxastic state should be such that its representation excludes all and only the degree of belief assignments that the evidence makes untenable, were it to allow a precise assignment of degrees of belief. In this case, coherence, and so evidence, rules out every nonprobabilistic degree of belief assignment. However, a requirement for assignments of degrees of belief in a hypothetical situation warranting precision does not clearly yield a requirement in the agent's actual situation. Indeed, the hypothetical situation is unintelligible when an agent's evidence, because sparse, prohibits a precise assignment of degrees of belief.

This subsection maintains that the requirement of coherence for a degree of belief assignment applies even when the assignment belongs to a set of multiple assignments representing an agent's doxastic state. Even in this context, a degree of belief assignment is incoherent if it fails to comply with the probability axioms. Incoherent degree of belief assignments (according to conventions for using them to represent a doxastic state) do not represent a coherent doxastic state. For a doxastic state to be coherent, an adequate representation must have coherent assignments of degrees of belief. If a degree of belief assignment in a doxastic state's representation is not a probability function, then the state that the set represents is incoherent.

In this way, requirements of coherence build an epistemic, noninstrumental, nonderivative, and explanatory argument that the degree of belief assignments in the representative set must be probability functions. The argument for the norm appeals to the intrinsic value, from doxastic rationality's perspective, of a coherent doxastic state that a set of probabilistic degree of belief assignments represents.

To confirm the norm, I argue that some of its implications are, indeed, coherence constraints on doxastic comparisons of propositions. I show that compositionality and separability – as Subsection 4.4 introduces these features of doxastic attitudes to exclusive propositions forming a disjunction – follow from the additivity of the degree of belief assignments, or the representatives, in a set of assignments representing a doxastic state.

Take compositionality first. If, according to each representative, the degree of belief that an exclusive disjunction holds equals the sum of the degrees of belief that its disjuncts hold, then the only factors affecting the doxastic attitude to the disjunction are the doxastic attitudes to the disjuncts. The representatives, and thus the representation of the agent's doxastic state, reveal no other factors relevant to rational choice. Therefore, the attitude to the disjunction is a function of the attitudes to the disjuncts, as compositionality requires.

Similarly, if, according to each representative, the degree of belief that an exclusive disjunction holds is the sum of the degrees of belief that its disjuncts hold, then the disjuncts are separable from each other. Each representative, and thus the representation, shows agreement between the order of first disjuncts and the order of disjunctions when exclusive disjunctions have a common second disjunct. Also, they show agreement between the order of second disjuncts and the order of disjunctions when exclusive disjunctions have a common first disjunct.

As Subsection 4.4 argues, the independence of the reasons for doxastic attitudes to exclusive propositions supports the compositionality and separability of doxastic comparisons. These features of doxastic comparisons are implications of representatives' complying with the probability axioms. That norms of doxastic comparison follow from probabilistic representatives confirms, as a requirement of rationality for ideal agents, that representatives comply with the probability axioms. Just as the norm of compliance with the axioms of probability applies to a single assignment of degrees of belief that represents a quantitative doxastic state, it applies to each assignment of degrees of belief in a set of assignments that represents a nonquantitative doxastic state.

6.2 Imprecise Utilities

If an agent has little experience but complete information, a set of degree of desire assignments by itself may represent her conative state. If she also has incomplete information, a representation of her conative state pairs a degree of belief assignment with each degree of desire assignment because her conative state depends on her doxastic state.

In a decision problem, a representation of an agent's doxastic and conative state uses a set of pairs of a degree of belief assignment to possible outcomes of options and a degree of desire assignment to possible outcomes of options and also to options. Coherence requires that each degree of belief assignment be a probability function and also that each degree of desire assignment be a utility function. That is, it requires that each pair, or representative, assign a degree of desire to an option that equals the expected degree of desire of the option's outcome, so that the degree of desire assigned to the option complies with the expected-utility principle. The coherence requirements for degrees of desire, as for degrees of belief, apply even in the context of a representation of a doxastic and conative state by a set of pairs of a degree of belief assignment and a degree of desire assignment.

A confirmation of the expected-utility principle for a representation's degree of desire assignments shows that the principle implies the norm of transitivity for an agent's preferences among options. Transitivity requires that, for options o_1, o_2, and o_3, if an agent prefers o_1 to o_2, and prefers o_2 to o_3, then she also prefers o_1 to o_3. A representative's degree of desire assignment compares options according to the degrees of desire it assigns to the options. If the representative complies with the expected-utility principle, these comparisons of options are the same as comparisons of options according to expected degrees of desire computed using the representative's degree of belief assignment and degree of desire assignment and therefore are transitive. An agent's preferences among options agree with comparisons of options that the representatives have in common and so are transitive too.

6.3 Norms Given Imprecision

As Subsection 2.5 explained, norms for a doxastic state, a conative state, and a combination of a doxastic and conative state apply to the states via their representations, given that the representations are adequate and so depict all the features of a state that matter to its evaluation for rationality. For an ideal agent, rationality requires a degree of belief assignment to be a probability function and a degree of desire assignment to be a utility function, no matter whether the assignments belong to a representation of an agent's quantitative or nonquantitative doxastic and conative state. Although requirements for a state's representation derive from requirements for the state, requirements for the representation reveal requirements for the state. The state must be such that its representation meets coherence requirements.

Furthermore, norms for nonquantitative doxastic and conative states derive from norms for quantitative doxastic and conative states because norms for nonquantitative states apply via their representations using pairs of a degree of belief assignment and a degree of desire assignment. For example, if a norm for a quantitative doxastic state entails a norm for degrees of belief, then the norm for degrees of belief also entails a norm for a nonquantitative doxastic state. Norms for a quantitative doxastic state, through norms for degrees of belief, ground norms for a nonquantitative doxastic state.

The norms for attitudes derive from the norms for states containing them. For example, the norms for a doxastic state imply norms for the doxastic attitudes that the state contains. Suppose that a set of multiple assignments of degrees of belief represents a doxastic state with doxastic attitudes to p and to $(p \lor q)$. Each assignment in the set, through its values for p and for $(p \lor q)$, may represent a doxastic relation between the two propositions, although the doxastic state's representation does not assign a single degree of belief to p and a single degree of belief to $(p \lor q)$. According to a comparative norm, belief that p should be no greater than belief that $(p \vee q)$. An expression of this comparative norm does not need representations of the doxastic attitudes to p and to $(p \vee q)$. A version of the norm may use the degree of belief assignments in the set representing the doxastic state containing the attitudes. Expressed using these assignments, the norm requires that, for every assignment in the set, the degree of belief that p is no greater than the degree of belief that $(p \vee q)$; conformity to a law of probability for entailments requires that $B(p) \leq B(p \vee q)$. This norm for a degree of belief assignment reveals a norm for a nonquantitative doxastic state, with doxastic attitudes to p and to $(p \vee q)$, requiring that belief that p be no stronger than belief that $(p \vee q)$. In general, norms for degree of belief assignments representing a doxastic state reveal norms for the state and doxastic attitudes the state contains.

A representation of an agent's doxastic attitude to p may be a set of degrees of belief, and similarly for $(p \vee q)$. However, the sets for the two attitudes miss the ground of the comparative norm – namely, for p and for $(p \vee q)$, the relation between the values of their degree of belief assignments in each assignment of the set of degree of belief assignments that represents the doxastic state containing the doxastic attitudes. Two sets of degree of belief assignments, one for each proposition, do not reveal the comparative norm that governs the attitudes to the two propositions.

An agent complies with the comparative norm concerning p and $(p \vee q)$ if the set that represents her doxastic state has degree of belief assignments that are probability functions. If all are probability functions, then together they imply that she does not believe p more strongly than she believes $(p \vee q)$, because her doxastic state includes comparisons entailed by all the degree of belief assignments in the set representing her doxastic state. Rationality constrains non-quantitative doxastic attitudes by constraining the degree of belief assignments that represent a nonquantitative doxastic state containing the doxastic attitudes. Nonquantitative doxastic attitudes are rational only if, in the set of degree of belief assignments representing the doxastic state with the attitudes, each assignment is a probability function.

In general, because a set of pairs of a degree of belief assignment and a degree of desire assignment represents a doxastic and conative state, rationality's constraints on the pairs of assignments reveal its constraints on the doxastic attitudes and conative attitudes in the state. Laws of probability and utility governing the pairs of assignments reveal constraints on the doxastic and conative attitudes in the state. The attitudes must conform to the consequences of the pairs' complying with the laws of probability and utility. Hence, given that, for every pair, some act in a decision problem has greater expected utility than another act, a rational ideal agent wants to perform the first act more strongly than the second.

Norms of rationality for an agent's imprecise doxastic and conative attitudes derive from norms for the doxastic and conative state to which these attitudes belong. Moreover, norms of coherence for representations of the state reveal norms of coherence for the state and the attitudes.

7 The Permissive Principle of Choice

Rationality's requirements take account of an agent's circumstances. Consequently, an agent may make a rational choice despite having little relevant evidence and experience and, therefore, without having precise probabilities and utilities for possible outcomes of options.[40] For example, a student may make a rational choice about entering medical school despite possible outcomes having imprecise probabilities and utilities.[41]

This section argues for the permissive principle of choice that Section 2 assumes when characterizing a representation of an agent's doxastic and conative state minimally adequate for a decision problem she faces. The argument assumes the decision model that Subsection 2.1 constructs, including its idealizations about agents, their decision problems, and their situations.

In the decision model, an agent is ideal and is rational, except perhaps in the current decision problem, and is in ideal conditions for complying with the permissive principle of choice. A set of pairs of a degree of belief assignment and a degree of desire assignment represents the agent's doxastic and conative state, and the state accords with the agent's evidence and experience. Because the agent is ideal and rational, the elements of a pair are, respectively, a probability function and a utility function, so, to simplify, I call the degree of belief assignment a probability assignment and call the degree of desire assignment a utility assignment.

The permissive principle of choice for an agent with imprecise probability and utility assignments requires a decision to maximize expected utility according to some *representative*, that is, a pair of a probability assignment and a utility assignment from the set of pairs representing the agent's doxastic and conative state. I call the principle permissive because a typical rival principle prohibits more options than it does.[42] However, the agent's doxastic and conative state includes any internal reason for prohibiting an option, the type

---

[40] *Pace* Paul (2014).

[41] Bradley (2015) surveys decision principles that use imprecise probabilities.

[42] Bradley and Steele (2016: sec. 4) adopt a more permissive decision principle that prohibits an act by an agent if and only if, for some alternative act, it is inferior to the alternative act given every probability assignment in the set of probability assignments representing the agent's doxastic state. Their principle permits any choice that the permissive decision principle permits. However, it also may permit a choice that fails to maximize expected utility according to any representative and so according to every representative is inferior to some other choice. For

of reason rationality weighs. Hence, all such reasons influence the agent's set of representatives and through them restrict the set of permissible options as much as reasons warrant. For example, a representative's utility assignments to possible outcomes register aversion to an option's risk. Hence, an agent's aversion to risk does not prohibit any option that maximizes expected utility according to a representative. Such a prohibition would double-count aversion to risk.[43]

Ellsberg's (1961) paradox presents preferences that exhibit aversion to uncertainty, or, as the literature says, *ambiguity*, about the probabilities of an act's possible outcomes. In one sense, ambiguity arises if the agent does not know the physical probabilities of the act's possible outcomes. However, in another sense, which I adopt, it arises if the agent's assignment of evidential probabilities to the act's possible outcomes are imprecise because of scant evidence. Ignorance of the physical probabilities of options' possible outcomes does not entail ambiguity in this second sense, for an agent's evidential probabilities may be precise despite such ignorance.

In a simple version of Ellsberg's paradox, an agent forms a preference between (1) a gamble that pays $100 if a draw produces red from an urn with 50 red balls and 50 green balls and (2) a gamble that pays $100 if a draw produces red from an urn with an unknown mixture of red and green balls. Also, the agent forms a preference between (3) a gamble that pays $100 if a draw produces green from the first urn and (4) a gamble that pays $100 if a draw produces green from the second urn. A typical agent prefers (1) and (3), the gambles involving the urn with the known mixture of red and green balls. Taking these preferences to be rational, they may seem to challenge the permissive principle of choice, for according to no representative do both preferences follow expected utilities given that possible outcomes are just amounts of money. In this case, if the preferences follow expected utilities, then, according to the first preference, the probability of red is greater from the first urn than from the second; and, according to the second preference, the probability of green is greater from the first urn than from the second and, hence, contrary to

example, suppose that about a coin an agent knows only that it has either two heads or has two tails, and so concerning a toss of the coin is in a doxastic state represented by two probability assignments, one with $P(\text{heads}) = 1$ and the other with $P(\text{heads}) = 0$. Imagine that the agent's options are getting $1 if a toss of the coin yields heads, getting $1 if the toss yields tails, and getting nothing. The extremely permissive principle permits picking nothing because no alternative is better according to each representative. Getting $1 if heads is not better if $P(\text{heads}) = 0$, and getting $1 if tails is not better if $P(\text{heads}) = 1$. However, picking nothing is an irrational choice.

[43] Bales, Cohen, and Handfield (2014), Schoenfield (2014), and Bradley (2015) argue that in some cases the permissive principle allows an irrational choice. Their objections depend on controversial standards of rationality.

the first probability comparison, the probability of red is less from the first urn than from the second.

Ambiguity aversion, which I assume is rational, is aversion to a type of risk an option in a decision problem generates given insufficient evidence for probability assignments to the option's possible outcomes.[44] An agent's attitude to an option's risk, taken as a consequence of the option, comprehends the agent's attitude to ambiguity, as ambiguity may be a factor in an option's risk. An option's possible outcomes, by including the option's risk, include any ambiguity the option creates. Because the agent's evaluation of a possible outcome of an option comprehends ambiguity, an agent averse to ambiguity may rationally have the preferences that are typical in Ellsberg's paradox. The agent's preferences follow expected utilities according to a representative if possible outcomes include ambiguity and the agent is sufficiently averse to ambiguity.

In Ellsberg's paradox, that a gamble brings ambiguity is a consequence of the gamble. Rationality imposes few restrictions on an agent's attitude to the consequence. Suppose that an agent with linear utility for money cares only about ambiguity and chances for money. Rationality allows a range of prices to be the lowest at which she is willing to exchange gamble (1) for gamble (2) and allows a similar range for an exchange of gamble (3) for gamble (4). Rationality does not strictly regulate the strength of her aversion to ambiguity.

Because rationality is tolerant concerning an agent's attitude to ambiguity, a decision principle cannot successfully exclude ambiguity from the possible outcomes of options and then mandate a general adjustment of an option's utility to make up for ambiguity's exclusion. The permissive principle accommodates rationality's tolerance concerning an agent's attitude to ambiguity by letting possible outcomes of an option include ambiguity about the option's outcome and by letting an agent evaluate possible outcomes according to her own lights.

In the decision model adopted, the permissive principle of choice formulates a necessary and sufficient condition of a rational decision, namely maximizing expected utility according to a representative, a pair of probability assignment and a utility assignment in the set representing the agent's doxastic and conative state. The model's idealizations about agents put aside restricted access to probability and utility assignments and other factors that generate excuses for not satisfying the permissive principle. Its idealizations about decision problems put aside problem cases, such as those in which no option stably

---

[44] Al-Najjar and Weinstein (2009) review the literature on aversion to ambiguity and argue that such aversion is irrational. Siniscalchi (2009) defends its rationality.

maximizes expected utility given any representative. Because the permissive principle accommodates every consideration bearing on choice-worthiness, satisfying it is sufficient as well as necessary for a rational choice.

A rational ideal agent, because of limited evidence and experience, may have an incomplete preference-ranking of options. The permissive principle says for every option whether it is rational but it need not generate a preference-comparison for every pair of options. It may leave two options unranked.

A general principle of choice, using preferences among options instead of probabilities and utilities of possible outcomes of options, requires an agent to realize an option such that no option is preferred, that is, *a top-ranked option*, and takes meeting this requirement as the standard for a rational choice. Suppose that an agent who must choose between hiking and listening to music does not compare these options and so has no preference between them and yet is not indifferent between them. In this case, rationality permits each option. Each is top-ranked.

The general principle of choice supports the permissive principle because rational preferences among options agree with the expected utilities of options according to a representative. An agent prefers one option to another if and only if according to every representative the first option has greater expected utility than has the second. Given this relation between preferences and expected utilities according to a representative, an option is top-ranked if and only if it maximizes expected utility according to some representative. Satisfying the permissive principle is equivalent to realizing a top-ranked option. Thus, the permissive principle has the support of the general principle of choice.

Joyce (2010) objects to the permissive principle, which he calls the liberal principle. For bets, he characterizes the principle as follows: "It is mandatory to take bets that uniquely maximize expected utility for all [representatives]. It is permissible to take any set of bets that maximizes expected utility relative to some [representative]" (p. 314). I interpret the permission to apply to a set of simultaneous bets, which in a decision problem the permissive principle treats as one option because an agent's decisions to make each bet amount to a decision to make all. Joyce holds that the permissive principle is too permissive and adds additional constraints on rational choice in decision problems with imprecise probabilities. However, the permissive principle yields rational choices without additional constraints; by itself it brings to bear every relevant consideration.

Another objection to the permissive decision principle is that a rational choice depends not just on doxastic and conative attitudes to possible outcome of options but also on attitudes to the risks that options generate, taking an option's risk in the sense of the variability of the utilities of its possible outcomes according to representatives. Some theorists, such as Gärdenfors and Sahlin (1982), assume that rationality requires a strong aversion to risk and so advance the principle to maximize minimum expected utility (MMEU) with respect to the set of pairs of a probability function and a utility function that represents an agent's doxastic and conative state. However, rationality tolerates diverse attitudes to an option's risk. Some rational agents are attracted to the risk an option generates; they may enjoy a gamble because of its risk. Rationality does not legislate a strong aversion to an option's risk. Because MMEU rests on the assumption that rationality requires a strong aversion to an option's risk, it is too strict a principle of choice.

Some arguments against the permissive principle of choice target its application to sequences of choices, claiming that a sequence of choices has irrational elements if it does not maximize expected utility among rival sequences of choices. However, standards of rationality for sequences of choices do not support this claim, as Section 8 argues.

8 Sequences of Choices

The permissive principle of choice governs all choices, including the choices in a sequence of choices. I examine its application to these choices because some criticisms of the principle claim that it permits irrational sequences of choices and therefore irrational choices.

This section maintains that the rationality of a sequence of choices depends on the rationality of its steps. An examination of quantitative cases supports this claim. In these cases, although a sequence's having rational steps and its maximizing utility among rival sequences generally go hand in hand, when they come apart, rationality evaluates a sequence by evaluating its steps.

Before considering evaluation of a sequence of choices, for background, consider evaluation of another type of combination of choices, namely choices made simultaneously. An evaluation treats adoption of a combination of simultaneous choices as a single choice because an agent can at will adopt the combination. A combination of decisions made all at the same time is rational only if it maximizes (expected) utility; and if it maximizes, then its components do. For example, a system of bets made all at the same time is rational only if the system maximizes utility. Furthermore, each bet in the system maximizes utility if the system does. In a maximizing system, each bet maximizes utility because it completes a system that maximizes utility. Given the other bets, declining the bet, if declining does not maximize, produces a system with less utility than the maximizing system that accepting produces.

Rationality evaluates a decision at a time using an evaluation of a full specification of the decision rather than a partial specification. To evaluate buying a hat, it evaluates all features of the purchase, including color selection. Buying a hat is rational only if the color selection maximizes given the other features of the purchase. A multifeatured act is rational only if each feature maximizes given the others. Similarly, a multicomponent decision maximizes only if each component maximizes given the other components. Consequently, a system of bets is rational only if each bet in the system maximizes given the other bets. Rationality does not evaluate in isolation a bet in a system of bets adopted at a time but evaluates the bet under the assumption of the other bets. The bet is rational only if it maximizes given the other bets.

In contrast, a combination of decisions made one after the other may maximize utility although some decisions in the sequence do not maximize utility given the other decisions. This happens in cases in which the sequence has consequences not tracible to any single decision in the sequence. For example, because of aversion to risk an agent may maximize utility by declining a gamble that pays $4 if heads turns up on a coin toss and loses $2 if tails turns up and then maximize utility by declining a gamble on a second coin toss with the same payoffs. Although each gamble has an expected payoff of $1, the agent may be rationally averse to risk and so rationally decline each gamble. However, accepting both gambles reduces the risk of a net loss and may reduce it to a point at which accepting both gambles maximizes utility. Accepting both gambles has the possible outcomes of gaining $8 with probability ¼, gaining $2 with probability ½, and losing $4 with probability ¼. The combination diminishes the probability of losing, although it increases the size of a possible loss. The agent may find the combination attractive because of the increased probability of a gain and because of the sizes of possible gains. Samuelson (1963) dramatizes the point by supposing a sequence of 100 offers of such bets. Declining each bet is rational because of aversion to risk but accepting all bets brings little risk and offers a high probability of a large gain. For an agent, the sequence of acceptances may maximize utility although it does not have utility maximizing steps.

A rational choice in a sequence considers the effect of its combination with the other choices in the sequence. In a sequence of choices about the two gambles concerning coin flips, an agent with an aversion to risk may see that she will reject the last gamble and so, applying backwards induction, see that she maximizes utility by rejecting the first gamble. Her rejection of each gamble in sequence is rational although the sequence of rejections does not maximize among rival sequences.

Because a maximizing sequence of choices need not have maximizing steps, rationality evaluates a sequence of choices not by comparing it with rival sequences but by evaluating its steps. If they are rational, then the sequence is rational. Rationality treats differently sequences of choices and simultaneous choices because an agent has direct control over present acts only. At a time, an agent can at will perform only acts at the time. Rationality does not evaluate a sequence of choices using utility maximization because an agent cannot perform the sequence at will. The agent can at will realize only each step in the sequence at the time for it. Because of this limit on control, rationality evaluates a sequence by evaluating its steps. If each step is rational, and so maximizes utility, then the sequence is rational, even if the sequence does not maximize utility.[45]

If a sequence of rational choices produces a non-maximizing sequence, as in the case of the sequence of offers to bet on coin flips, a rational agent, given the opportunity, may bind herself to performing acts that yield a maximizing sequence. Binding may operate in various ways. The binding may occur by making a resolution she abhors breaking, by entering a contract to perform the acts, or by setting up a penalty for deviation from the sequence. It may also occur by putting the choices in the sequence out of her hands and having another party make them on her behalf. If at the beginning of a sequence an agent has an opportunity to bind herself to a sequence of decisions, then that binding is an act that rationality evaluates using the standard of maximization of (expected) utility. The binding may maximize utility and so be rational. However, it is an act at a single time performable at will. Although it yields a sequence of acts, it is not a sequence of multiple acts.

An act's belonging to a sequence to which an agent rationally binds herself does not ensure the rationality of the act. Take Newcomb's problem, a sequence with just one choice, the choice to take an opaque box along with a transparent box that contains $1,000 or just the opaque box in the hope it contains $1,000,000 because of a prediction that only it is chosen. An agent may rationally bind herself, by taking a one-boxing pill before facing the problem, to picking only one box when facing the problem – taking the pill is likely to prompt a prediction that fills the opaque box with $1,000,000. Nonetheless, picking only the opaque box is irrational, I assume, because picking both boxes strictly dominates it. An agent's binding herself to an irrational act with bad consequences may be rational because of the good consequences of the binding in cases that reward irrational acts.[46]

---

[45] Having rational steps is a sufficient condition of a sequence's rationality but not a necessary condition because a rational sequence may contain irrational steps that are inconsequential given the whole sequence. A necessary and sufficient condition of a sequence's rationality is having the same utility as a sequence in which each step is rational. I state this necessary and sufficient condition for completeness but use only the sufficient condition to defend the permissive principle's application to the choices in a sequence.

[46] Gaifman (1983: 150–153) shows that rewarding irrationality may create a paradox of rationality structurally similar to the paradox of the liar, in which rationality has a role similar to truth's role in the paradox of the liar. Responses to the paradox of the liar generate similar responses to the paradox of rationality. However, I do not argue for any particular response because the paradox is too complex to address briefly.

In Newcomb's problem, binding can work by creating a powerful incentive to one-box. A rich agent may make an arrangement whereby if she takes two boxes, she forfeits more money than she might gain by taking two boxes. This type of binding makes one-boxing rational because it evades a penalty. The penalty makes it rational to perform an act that would otherwise be irrational. However, as mentioned, an agent may bind herself to one-boxing by taking a pill that causes her to one-box in Newcomb's problem (and has no other relevant effects). The pill causes her to perform an irrational act. This can happen although the agent still chooses to one-box, just as smoking-cessation pills may cause an agent to forego opportunities to smoke although she still chooses not to smoke.

Rationality's standards for each choice in a sequence of choices require that it evaluate the sequence by evaluating its steps. Rationality does not leave room for independent multi-chronic standards because its synchronic standards completely govern every choice. Rationality imposes no requirement of utility maximization on a sequence of choices that is independent of its requirements on the choices in the sequence. Rationality's multi-chronic requirements emerge from its synchronic requirements. This holds in nonquantitative cases as well as quantitative cases.

Some objections to the permissive principle claim that it may lead to an irrational sequence of choices. Before looking at cases, I first note that, during a sequence of choices, an agent's information and goals may change. These changes may make rational choices produce a sequence that does not maximize utility among alternative sequences.

For example, a shopper may decide to go to a new supermarket but en route decide to go instead to his usual supermarket. The change in goal wastes time but the sequence of choices is not thereby irrational. Changes of mind need not be irrational. Similarly, the shopper may decide to go to the new supermarket but en route see a sign announcing that its opening date is tomorrow and so go to his usual supermarket today. The new information lowers the utility of going to the new supermarket. Although the sequence of decisions wastes time, it is not irrational. Examples purporting to show that rational choices may yield an irrational sequence put aside such excuses for sequences that do not maximize utility by assuming that goals and relevant information are constant throughout the sequence. An agent's sequence of choices need not maximize utility if she does not have constant goals and constant relevant information throughout the sequence.

The assumption of constant relevant information needs relaxation for sequences of choices in which chance settles whether the agent arrives at a stage of the sequence. Reaching the stage brings information that the chance event occurred. Its occurrence may be relevant new information that excuses a sequence of choices' failing to maximize utility. In a sequence in which chance effects the choices the agent faces, applying utility maximization to the sequence assumes only that at each stage of the sequence no relevant information arrives that the agent did not at earlier stages anticipate having at that stage should she reach that stage.

Another preparatory point concerns the principle of dominance. In simultaneous choices, it applies to a system of choices because an evaluation of the system may treat the system as one choice. An extension of the principle asserts that it is irrational to realize a sequence of choices if another sequence dominates it. This is a multi-chronic principle because it governs choices at multiple times. However, the principle of dominance does not govern rationality's evaluation of a sequence of choices, just as the related principle of utility maximization does not govern the evaluation. Rationality's synchronic standards for evaluating each choice in a sequence leave no room for an independent multi-chronic standard for evaluating the whole sequence of choices; in particular, they leave no room for an extension of the principle of dominance to sequences of choices.

A typical objection to the permissive principle supposes that an agent has a ticket that pays a dollar if it rains tomorrow and nothing otherwise. For the agent, the probability of rain is imprecise and the probability functions that represent her doxastic state have a range of values for rain tomorrow going from a low of 0.4 to a high of 0.6. It seems that the permissive principle permits her to sell the ticket for 40 cents and then to buy it back for 60 cents, with a net loss of 20 cents. Moreover, this sequence of choices seems irrational because not making the two transactions is sure to be better; it dominates making the transactions.

A reply first notes that the multi-chronic principle of dominance is not authoritative for sequences of choices. A sequence is rational if its steps are rational. Second, an ideal agent does not make the dominated sequence of transactions. An ideal agent predicts the choices she will make in foreseen decision problems. She uses self-knowledge extending beyond knowledge of her own rationality to foresee her exercise of rationality's permissions. Granting that the agent knows she will have an opportunity to buy back the ticket if she sells it, she predicts whether she will buy it back. If she predicts that she will buy it back for 60 cents, then she will not sell it for 40 cents. Given her prediction, not selling maximizes utility using any probability function in the set representing her doxastic state. Foresight prevents an ideal agent from making the dominated sequence of transactions.

Elga (2010) presents the case of an agent Sally who is offered in sequence two gambles A and B, as in Figure 3. Gamble A pays \$15 if hypothesis h is false and costs \$10 otherwise. Gamble B pays \$15 if h is true and costs \$10

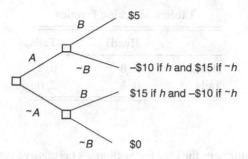

Figure 3 A sequence of offers

otherwise. By accepting both gambles, Sally guarantees a gain of $5. However, if a representation of Sally's doxastic state yields a probability interval [0.1, 0.8] for *h*, then the permissive principle allows her to reject each gamble, the objection claims. She may reject gamble *A* using 0.8 as *h*'s probability and reject gamble *B* using 0.1 as *h*'s probability. Because this sequence of choices is dominated by accepting both gambles, Elga rejects the permissive principle.

Again, a reply has two parts. First, the multi-chronic principle of dominance is not authoritative for evaluation of sequences; the rationality of a sequence's steps make the sequence rational. Second, as Weirich (2015b) explains, if Sally is an ideal agent who predicts her own choices she does not reject both gambles. If she predicts her rejecting gamble *B*, then she does not rationally reject gamble *A*; the only reason for rejecting it is preparation to accept gamble *B* and thereby stake her chances for money on *h*'s being true. If Sally rejects gamble *A* and then rejects gamble *B*, either rejecting gamble *B* is a mistake because it foregoes the opportunity to gain from gamble *B* or, if it is not a mistake, then rejecting gamble *A* is a mistake because it squanders the opportunity to guarantee $5. The exercise of a future permission to reject gamble *B* removes the current permission to reject gamble *A*. Rejecting gamble *A* is irrational given that if she rejects gamble *A*, she will reject gamble *B*. Rejecting gamble *A*, given that rejecting gamble *B* will follow, does not maximize utility according to any pair of a probability assignment and a utility assignment in the set of such pairs representing her doxastic and conative state. Hence Sally, if she is an ideal agent who complies with the permissive principle, does not reject gamble *A* and then reject gamble *B*.[47]

---

[47] In a sequential version of Allais's (1953) paradox, it may appear that the permissive principle allows choices that are intertemporally incoherent. However, Weirich (2018) argues that a change in information during the sequence of choices rebuts the charge of intertemporal incoherence. Hammond (1988) presents a case in which two apparently permissible choices,

Table 1 Matching Pennies

	Heads	Tails
Heads	2, 0	0, 2
Tails	0, 2	2, 0

This section buttresses the case for Section 7's permissive principle of choice by defending its application to the steps of a sequence of choices. The defense relies on stepwise principles for evaluating sequences of choices and also idealizations about agents. It shows the rationality of sequences of choices that comply with the permissive principle.

9 Choices in Games of Strategy

Decisions in games may be made using imprecise probabilities and utilities. Hence, a general theory of rational choice using imprecise probabilities and utilities covers choices in games. This section shows that imprecision undermines an argument for realization of a Nash equilibrium of a finite game of strategy with simultaneous moves and therefore motivates an account of solutions to games that does not require a Nash equilibrium.

A typical game of strategy with simultaneous moves is Matching Pennies. In this game, with two players, each player displays a penny with one side up, either heads or tails. One player wins both pennies if the sides up are the same and the other player wins both pennies if the sides up differ. Table 1 presents the game letting rows represent the strategies of the player, Row, who gains from a match and letting columns represent the strategies of the other player, Column. The cells contain the players' payoffs in pennies, giving first Row's payoff and second Column's payoff.

A table, or matrix, for a game represents explicitly each player's pure strategies. Beside pure strategies, the players may have mixed strategies that are probabilistic mixtures of pure strategies, such as the mixture that for a player assigns probability ½ to displaying heads and probability ½ to displaying tails – abbreviating, (½ Heads, ½ Tails). To implement this mixed strategy, a player might select a pure strategy by tossing a coin. An implementation binds the player to deciding among pure strategies according to the coin toss. Given that a mixed strategy may assign probability 1 to a pure strategy, the mixed strategies include the pure strategies.

---

at different nodes of the second stage of a decision tree, lead to a dominated sequence of choices. However, exercising one permission prohibits exercising the other permission.

A *Nash equilibrium* is a profile of strategies, a combination of strategies with exactly one strategy for each player, such that each strategy is a best response to the others. Matching Pennies has no equilibrium in pure strategies. However, it has an equilibrium in mixed strategies, namely the mixture (½ Heads, ½ Tails) for Row and the same mixture for Column. This mixed strategy equilibrium is the only equilibrium of the game.

A payoff matrix describes several, but not all, relevant features of a game. It does not describe the players' traits and beliefs about each other, for example. Additional assumptions fill in the other features relevant to a player's decision. In the decision model I adopt, the players are ideal and rational.

Suppose that the players understand the matrix that represents their game and predict each other's choices – in the case of a mixed strategy chosen, just the mixed strategy and not also the pure strategy it yields. Because of the players' predictive power, a player's choice furnishes evidence of the other player's choice, as the other player responds rationally to the player's choice given its prediction. In Matching Pennies, each player adopts the mixed strategy (½ Heads, ½ Tails) because the strategy profile with these mixed strategies is the only strategy profile compatible with the assumptions concerning the player's information and rationality. Given that Row predicts this mixed strategy by Column, Row's adopting the same mixed strategy maximizes utility and similarly for Column. Any other strategy profile, contains a strategy for a player that does not maximize utility for the player given the player's knowledge of the other player's strategy. For example, the profile in which Row adopts the mixed strategy (¼ Heads, ¾ Tails) and Column adopts the mixed strategy (½ Heads, ½ Tails) is not compatible with the assumptions because Column's best response to Row's mixed strategy is displaying heads and not the mixed strategy (½ Heads, ½ Tails). Under the assumptions, the players realize an equilibrium and, because in Matching Pennies only one equilibrium exists, they realize it.[48]

In a game of Matching Pennies, each player uses information about the other player to assign a probability to the other player's displaying heads. This information may include information about the other player's information about the first player's information and so on. However, even given the players' common knowledge of their game's payoff matrix and their rationality, each player's assignment of probabilities to the other player's strategies may be imprecise.[49]

---

[48] A Nash equilibrium exists in every game of strategy with a finite number of players and a finite number of pure strategies for each player. In other games of strategy, there is a case for other types of equilibrium, such as strategic equilibrium, as Weirich (1998) defines it.

[49] A group has *common knowledge* of a proposition if and only if all know the proposition, all know that all know it, and so on.

In Matching Pennies, if each player assigns probability ½ to the other player's displaying heads, then she has a reason to display heads with probability ½; doing this maximizes utility. However, neither player has a reason to assign probability ½ to the other player's displaying heads given only the details of the game that its matrix presents. Furthermore, given only common knowledge that the players are rational and understand their game as its matrix presents it, neither has grounds for assigning a precise probability to the other's displaying heads. In their game, their probability assignments may be imprecise, even after reflection on their situation. Without giving the players more information about their game and each other, reflection does not produce for a player, first, a probability assignment of ½ to the other player's displaying heads, next, a decision to display heads with probability ½, and then, granting that the other player has insight into the player's deliberations, confirmation of the other player's displaying heads with probability ½. A player's deliberations do not have such dynamics because nothing in the game's payoff matrix produces an initial precise probability assignment to the other player's displaying heads. In particular, the notorious principle of indifference, I assume, does not warrant a probability assignment of ½.

In a game of strategy, only an equilibrium is compatible with the players predicting their choices and maximizing utility. A player, given her adoption of a strategy, assigns precise probabilities to the strategies of other players and rationality requires maximizing utility. However, when players do not predict precisely, and sets of probability assignments represent their doxastic states, many strategy profiles are compatible with the players' choosing rationally, that is, complying with the permissive principle of choice that Section 7 advances. Each player need only maximize utility according to some assignment in the set representing her doxastic state.

In Matching Pennies, consider the strategy profile in which each player adopts the mixed strategy (0.51 Heads, 0.49 Tails). This profile's realization is compatible with each player's maximizing utility according to a probability assignment in the set representing her doxastic state if, given that she adopts the mixed strategy, she assigns a range of probabilities that includes 0.5, such as the range from 0.4 to 0.6, to her opponent's displaying heads. This range may come from a range of mixed strategies for her opponent but may also come from uncertainty generated by scant evidence. Given that Row assigns such a range of probabilities to Column's displaying heads, all of Row's strategies maximize utility according to an element of the range, namely an assignment that gives probability 0.5 to her opponent's displaying heads. Plainly, Row's unique best response to (0.4 Heads, 0.6 Tails) by Column is (0 Heads, 1.0 Tails), and her unique best response to (0.6 Heads, 0.4 Tails) by Column is (1.0 Heads, 0 Tails). However, given (0.5 Heads, 0.5 Tails) by Column, all of Row's strategies, including the mixed strategy (0.51 Heads, 0.49 Tails), maximize utility. Similar results hold for Column. Hence, rational players need not realize an equilibrium.

In general, a player's having, given adoption of a strategy, imprecise probabilities for the strategies of the other players undermines the case for realization of an equilibrium. Rational players may not realize an equilibrium. Given imprecise probabilities and utilities, epistemic game theory identifies solutions to games that are not equilibria, understanding a solution as a profile of strategies such that each strategy is rational given the profile.[50]

10 Conclusion

A general account of rational choice in a decision problem accommodates an agent's lack of information and experience. It does not assume that, despite being in the dark, the agent assigns precise degrees of belief and precise degrees of desire to the possible outcomes of options. It accommodates imprecision.

Given ample information and experience, a rational ideal agent in a decision problem assigns degrees of belief to the possible outcomes of options and degrees of desire to the possible outcomes and also to the options themselves. The degrees of belief comply with the probability axioms and the degrees of desire comply with the principle of expected utility, thus forming, respectively, probabilities and utilities. Given sparse information and experience, a set of pairs of an assignment of degrees of belief and an assignment of degrees of desire represent an agent's doxastic and conative state. Rationality requires the assignments of degrees of belief to be probability functions and the assignments of degrees of desire to be utility functions. Hence, a set of pairs of a probability function and a utility function represent a rational ideal agent's nonquantitative doxastic and conative state. The representation's set of probability functions represent imprecise probabilities and its set of utility functions represent imprecise utilities.

I adopt a decision model with idealizations about an agent's abilities and circumstances. Under the model's assumptions, in a decision problem with precise probabilities and utilities for possible outcomes of options, an option is rational if and only if it maximizes expected utility. In a decision problem with imprecise probabilities and utilities, an option is rational if and only if it

---

[50] Perea (2012) offers an excellent introduction to epistemic game theory.

maximizes expected utility according to a pair of a probability function and a utility function in the set of pairs that represents the agent's doxastic and conative state. Rationality's standard for a choice made using imprecise probabilities and utilities is a generalization of its standard for a choice made using precise probabilities and utilities.

The standard of rationality for choices made using imprecise probabilities and utilities is permissive but no more permissive than reasons for choices warrant. It generates standards of rationality for sequences of choices and grounds epistemic game theory's exploration of solutions to games of strategy. Although not a demanding standard, it builds a strong foundation for a general theory of rational choice.

References

Allais, M. 1953. "Le comportement de l'homme rationnel devant le risque: Critique des postulats et axioms de l'école Américaine." *Econometrica* 21: 503–546.

Al-Najjar, N. and J. Weinstein. 2009. "The Ambiguity Aversion Literature: A Critical Assessment." *Economics and Philosophy* 25: 249–284.

Armendt, B. 1992. "Dutch Strategies for Diachronic Rules: When Believers See the Sure Loss Coming." In D. Hull, M. Forbes, and K. Okruhlik, eds., *PSA: Proceedings of the Biennial Meeting of the Philosophy of Science Association*, Vol. 1, pp. 217–229. Chicago: University of Chicago Press.

Augustin, T., F. Coolen, G. de Cooman, and M. Troffaes. 2014. *Introduction to Imprecise Probabilities*. Hoboken, NJ: Wiley.

Baccelli, J. and P. Mongin. 2020. "Can Redescriptions of Outcomes Salvage the Axioms of Decision Theory." Manuscript available at SSRN: ssrn.com /abstract=3610869ordx.doi.org/10.2139/ssrn.610869

Bales, A., D. Cohen, and T. Handfield. 2014. "Decision Theory for Agents with Incomplete Preferences." *Australasian Journal of Philosophy* 92: 453–470.

Bradley, S. 2015. "How to Choose among Choice Functions." Paper presented at the 9th International Symposium on Imprecise Probability: Theories and Applications, Pescara, Italy, July 20–24.

Bradley, S. 2019. "Imprecise Probabilities." In E. Zalta, ed., *Stanford Encyclopedia of Philosophy*. https://plato.stanford.edu/archives/fall2018/ entries/fundamentality/

Bradley, S. and K. Steele. 2016. "Can Free Evidence Be Bad? Value of Information for the Imprecise Probabilist." *Philosophy of Science* 83: 1–28.

Buchak, L. 2013. *Risk and Rationality*. Oxford: Oxford University Press.

Carnap, R. 1962. *Logical Foundations of Probability*. 2nd ed. Chicago: University of Chicago Press.

Castro, C. and C. Hart. 2019. "The Imprecise Impermissivist's Dilemma." *Synthese* 196: 1623–1640.

Chang, R., ed. 1997. *Incommensurability, Incomparability, and Practical Reason*. Cambridge, MA: Harvard University Press.

Christensen, D. 2004. *Putting Logic in Its Place*. Oxford: Oxford University Press.

de Finetti, B. [1937] 1964. "Foresight: Its Logical Laws, Its Subjective Sources." In H. Kyburg and H. Smokler, eds., *Studies in Subjective Probability*, pp. 93–158. New York: Wiley.

Easwaran, K. and B. Fitelson. 2012. "An 'Evidentialist' Worry about Joyce's Argument for Probabilism." *Dialectica* 66: 425–433.

Elga, A. 2010. "Subjective Probabilities Should Be Sharp." *Philosophers' Imprint* 10(5): 1–11. www.philosophersimprint.org/010005/

Ellsberg, D. 1961. "Risk, Ambiguity, and the Savage Axioms." *Quarterly Journal of Economics* 75: 643–669.

Evans, G. 1979. "Reference and Contingency." *The Monist* 62: 161–189.

Gaifman, H. 1983. "Paradoxes of Infinity and Self-Applications, I." *Erkenntnis* 20: 131–155.

Gärdenfors, P. and N.-E. Sahlin. 1982. "Unreliable Probabilities, Risk Taking, and Decision Making." *Synthese* 53: 361–386.

Good, I. J. 1952. "Rational Decisions." *Journal of the Royal Statistical Society, Series B*, 14: 107–114.

Greco, D. and B. Hedden. 2016. "Uniqueness and Metaepistemology." *Journal of Philosophy* 113: 365–395.

Hammond, P. 1988. "Orderly Decision Theory: A Comment on Professor Seidenfeld." *Economics and Philosophy* 4: 292–297.

Hart, C. and M. Titelbaum. 2015. "Intuitive Dilation." *Thought* 4: 252–262. doi.org/10.1002/tht3.185

Hedden, B. 2015. "Time-Slice Rationality." *Mind* 124: 449–491.

Howson, C. and P. Urbach. 2006. *Scientific Reasoning: The Bayesian Approach*. 3rd ed. Chicago: Open Court.

Jackson, E. and M. Turnbull. Forthcoming. "Permissivism, Underdetermination, and Evidence." In M. Lasonen-Aarnio and C. Littlejohn, eds., *Routledge Handbook for the Philosophy of Evidence*. Abingdon: Routledge.

Jeffrey, R. [1965] 1990. *The Logic of Decision*. 2nd ed., paperback. Chicago: University of Chicago Press.

Joyce, J. 1998. "A Nonpragmatic Vindication of Probabilism." *Philosophy of Science* 65: 575–603.

Joyce, J. 2010. "A Defense of Imprecise Credences in Inference and Decision Making." *Philosophical Perspectives*, 24: 281–323.

Konek, J. and B. Levinstein. 2019. "The Foundations of Epistemic Decision Theory." *Mind* 128: 69–107.

Krantz, D., R. D. Luce, P. Suppes, and A. Tversky. 1971. *Foundations of Measurement, Vol. 1: Additive and Polynomial Representations*. New York: Academic Press.

Lassiter, D. 2020. "Representing Credal Imprecision: From Sets of Measures to Hierarchical Bayesian Models." *Philosophical Studies* 177: 1463–1485.

Levi, I. 1974. "On Indeterminate Probabilities." *Journal of Philosophy* 71: 391–418.

Levi, I. 1980. *The Enterprise of Knowledge: An Essay on Knowledge, Credal Probability, and Chance.*Cambridge, MA: MIT Press.

Mahtani, A. 2019. "Imprecise Probabilities." In R. Pettigrew and J. Weisberg, eds., *Open Handbook of Formal Epistemology*, pp. 107–130. PhilPapers Foundation. https://philpapers.org/archive/PETTOH-2.pdf

Meacham, C. and J. Weisberg. 2011. "Representation Theorems and the Foundations of Decision Theory." *Australasian Journal of Philosophy* 89: 641–663.

Paul, L. A. 2014. *Transformative Experience*. Oxford: Oxford University Press.

Pedersen, A. P. and G. Wheeler. 2014. "Demystifying Dilation." *Erkenntnis* 79: 1305–1342.

Perea, A. 2012. *Epistemic Game Theory*. Cambridge: Cambridge University Press.

Ramsey, F. P. 1931. "Truth and Probability." In R. B. Braithwaite, ed., *The Foundations of Mathematics and other Logic Essays*, pp. 156–198. New York: Harcourt, Brace and Company.

Samuelson, P. 1963. "Risk and Uncertainty: A Fallacy of Large Numbers." *Scientia* 98: 108–113.

Savage, L. J. [1954] 1972. *The Foundations of Statistics*. 2nd ed.New York: Dover.

Schoenfield, M. 2014. "Decision Making in the Face of Parity." *Philosophical Perspectives* 28: 263–277.

Schoenfield, M. 2017. "The Accuracy and Rationality of Imprecise Credences." *Noûs* 51: 667–685.

Seidenfeld, T., M. Schervish, and J. Kadane. 2010. "Coherent Choice Functions under Uncertainty." *Synthese* 172: 157–176.

Seidenfeld, T., M. Schervish, and J. Kadane. 2012. "Forecasting with Imprecise Probabilities." *International Journal of Approximate Reasoning* 53: 1248–1261.

Shafer, G. 1976. *A Mathematical Theory of Evidence*. Princeton, NJ: Princeton University Press.

Siniscalchi, M. 2009. "Two out of Three Ain't Bad: A Comment on 'The Ambiguity Aversion Literature: A Critical Assessment'." *Economics and Philosophy* 25: 335–356.

Skyrms, B. 1987. "Coherence." In N. Rescher, ed., *Scientific Inquiry in Philosophical Perspective*, pp. 225–42. Pittsburgh: University of Pittsburgh Press.

Titelbaum, M. Forthcoming. *Fundamentals of Bayesian Epistemology*. New York: Oxford University Press.

Troffaes, M. and G. de Cooman. 2014. *Lower Previsions*. Hoboken, NJ: Wiley.

Walley, P. 1991. *Statistical Reasoning with Imprecise Probabilities*. London: Chapman and Hall.

Weirich, P. 1998. *Equilibrium and Rationality: Game Theory Revised by Decision Rules*. Cambridge: Cambridge University Press.

Weirich, P. 2001. *Decision Space: Multidimensional Utility Analysis*. Cambridge: Cambridge University Press.

Weirich, P. 2004. *Realistic Decision Theory: Rules for Nonideal Agents in Nonideal Circumstances*. New York: Oxford University Press.

Weirich, P. 2012. "Calibration." In H. de Regt, S. Hartmann, and S. Okasha, eds., *EPSA Philosophy of Science: Amsterdam 2009*, pp. 415–425. Dordrecht: Springer. doi.org/10.1007/978-94-007-2404-4_34/

Weirich, P. 2015a. *Models of Decision-Making: Simplifying Choices*. Cambridge: Cambridge University Press.

Weirich, P. 2015b. "Decisions without Sharp Probabilities." *Philosophia Scientiae* 19: 213–225.

Weirich, P. 2018. "Rational Plans." In J. L. Bermudez, ed., *Self-Control, Decision Theory, and Rationality*, pp. 72–95. Cambridge: Cambridge University Press.

Weirich, P. 2020. *Rational Responses to Risks*. New York: Oxford University Press.

Weisberg, J. 2009. "Commutativity or Holism? A Dilemma for Conditionalizers." *British Journal for the Philosophy of Science* 60: 793–812.

White, R. 2010. "Evidential Symmetry and Mushy Credence." In T. Szabo Gendler and J. Hawthorne, eds., *Oxford Studies in Epistemology*, Vol. 3, pp. 161–186. Oxford: Oxford University Press.

Zaffalon, M. and E. Miranda. 2018. "Desirability Foundations of Robust Rational Decision Making." *Synthese*: 1–42. https://doi.org/10.1007/s11229-018-02010-x

Zynda, L. 2000. "Representation Theorems and Realism About Degrees of Belief." *Philosophy of Science* 67: 45–69.

Acknowledgments

I thank Martin Peterson for encouraging me to present my views on the Element's topic, two anonymous reviewers for valuable suggestions, and Cambridge's team for excellent assistance.

For Rêva and Maurice.

Cambridge Elements ≡

Decision Theory and Philosophy

Martin Peterson
Texas A&M University

Martin Peterson is Professor of Philosophy and Sue and Harry E. Bovay Professor of the History and Ethics of Professional Engineering at Texas A&M University. He is the author of four books and one edited collection, as well as many articles on decision theory, ethics and philosophy of science.

About the Series

This Cambridge Elements series offers an extensive overview of decision theory in its many and varied forms. Distinguished authors provide an up-to-date summary of the results of current research in their fields and give their own take on what they believe are the most significant debates influencing research, drawing original conclusions.

A full series listing is available at: www.cambridge.org/EDTP

Cambridge Elements ≡

Decision Theory and Philosophy

Elements in the Series

Printed in the United States
By Bookmasters